Lecture Notes in Mathematics

Edited by A. Dold and B. Eckmann

955

Gerhard Gierz

Bundles of Topological Vector Spaces and Their Duality

Springer-Verlag
Berlin Heidelberg New York 1982

Author
Gerhard Gierz
Department of Mathematics, University of California
Riverside, CA 92521, USA

AMS Subject Classifications (1980): 46E10, 46E15, 46E40, 46H25, 46B20, 55R25, 28C20

ISBN 3-540-11610-9 Springer-Verlag Berlin Heidelberg New York
ISBN 0-387-11610-9 Springer-Verlag New York Heidelberg Berlin

Printed in Germany

Printing and binding: Beltz Offsetdruck, Hemsbach/Bergstr.
2146/3140-543210

Math Sci Sep

Contents

Introduction.

In the present notes we are dealing with topological vector spaces which vary continuously over a topological space. Among the first authors formulating this idea were Godement [Go 49], Kaplansky [Ka 51], Gelfand and Naimark. In these early papers, they axiomatized the idea of subdirect continuous representation of Banach spaces. To be precise, they considered spaces E of functions σ defined on a topological space X with values in given Banach spaces E_x, $x \in X$, satisfying axioms like

(1) The function $x \to \|\sigma(x)\| : X \to \mathbb{R}$ is (upper semi-) continuous and bounded for every $\sigma \in E$.

(2) The space E is complete in the norm $\|\sigma\| = \sup_{x \in X} \|\sigma(x)\|$.

(3) $E_x = \{\sigma(x) : \sigma \in E\}$ for every $x \in X$.

(4) E is a $C_b(X)$-module relative to the multiplication $(f,\sigma) \to f\cdot\sigma$: $C_b(X)\times E \to E$, where $(f\cdot\sigma)(x) := f(x)\cdot\sigma(x)$ and where $C_b(X)$ denotes the algebra of all continuous and bounded scalar valued functions on X.

In the following years, L. Nachbin, S. Machado and J.B.Prolla gave a similar definition for locally convex spaces. They, however, started from an approximation theoretical point of view.

The authors mentioned above were very well aware of the fact that their notion of continuous decomposition was closely related to continuous sections in fibre bundles. In fact, Fell [Fe 61], Dixmier and Douady [DD 63] as well as Dauns and Hofmann [DH 68] succeeded in giving a fibre bundle theoretical formulation of the axioms (1) - (4).

A third interesting aspect of spaces satisfying axioms (1) - (4) has its origin in sheaf theory and intuitionistic logic (see the Lecture Notes in Mathematics 753).

Originally, the work on the present notes was begun in order to give a useful description of the space of all continuous linear functionals on a space of sections in a bundle (or, equivalently, on a topological vector space satisfying axioms (1) - (4)). However, I have to admit that I did not succeed to my own satisfaction. Let me explain what my intention was at the beginning:

The best known example of a space satisfying axioms (1) - (4) is the space $C_b(X)$ of all continuous and bounded real-valued functions itself. If we assume for a moment that the base space X is compact, then the dual space of C(X) consists of all finite regular Borel measures on X acting on C(X) by integration. How should this fact generalize to a vector space E satisfying axioms (1) - (4)? Suppose that we start with a family $\Phi = (\phi_x)_{x \in X}$, where each ϕ_x is a continuous linear functional on E_x. Then every $\sigma \in E$ defines a real-valued mapping $\Phi(\sigma)$ by $\Phi(\sigma)(x) := \phi_x(\sigma(x))$. Suppose moreover, that μ is a measure on X and that $\Phi(\sigma)$ is bounded and μ-integrable for every $\sigma \in E$. Then we may define a linear functional ϕ on E by

$$\phi(\sigma) := \int_X \Phi(\sigma) \cdot d\mu .$$

Now the following questions arise:

1) Under which conditions on the family $(\phi_x)_{x \in X}$ is the function $\Phi(\sigma)$ integrable with respect to μ for every $\sigma \in E$?

2) Is every continuous linear functional representable in the form

$$\phi(\sigma) = \int_X \Phi(\sigma)\cdot d\mu \ ?$$

3) If so, how does one add $\int_X \Phi(\sigma)\cdot d\mu$ and $\int_X \Psi(\sigma)\cdot d\nu$, i.e. how is the algebraic structure of the dual space E' of E reflected in this integral representation of linear functionals?

A first answer to problem 1 of course would be that the mapping $\Phi(\sigma)$ is Borel measurable for every $\sigma \in E$. In this case, the family $(\phi_x)_{x\in X}$ yields a (bounded) linear mapping from E into the vector space of all bounded Borel measurable functions on X, equipped with the supremum norm. It is easy to see that $\Phi(f\cdot\sigma) = f\cdot\Phi(\sigma)$ for every $f \in C(X)$ and every $\sigma \in E$, i.e. Φ is a C(X)-module homomorphism.
It is very tempting to postulate that $\Phi(\sigma)$ is even continuous for the following reason: The space C(X) is the space of sections in the most simple bundle which can be thought of. Therefore and because C(X) acts on every space of sections by multiplication, this space should play the rôle of the field of the real numbers in the category of all spaces of sections in bundles over X. In this case, we would obtain a representation of linear functionals $\phi : E \to \mathbb{R}$ by an "internal linear functional" $\Phi : E \to C(X)$, i.e. by a continuous C(X)-module homomorphism into the "internal field" C(X) and a measure μ on X, i.e. an "external" linear functional on C(X).
Unfortunately, examples show that the linear functionals represented in this way do not even form a linear subspace of the dual space E' and that the linear span of these functionals does not have to be dense in E'.
I am aware of the fact that mathematicians like Burden and Mulvey viewing bundles from a point of view of sheaves, toposes and logic do not agree with my choice of the "internal real numbers". Therefore, they will not be suprised that I was not able to carry out my program in full generality (nor am I now) and it is certainly worthwhile

to check to what extend a use of their internal real numbers would lead to better results.

Problem 2 is solvable provided that X is a compact metric space (see section 21). An example of R.Evans (FU Berlin) conversely shows that this problem has a positive solution if and only if every finite regular Borel measure on X admits a strong lifting.

A discussion of the third problem leads to tensor products over C(X). Indeed, under certain (strong) restrictions, the dual space E' of E may be identified with a certain tensor product over C(X) between the space M(X) of all finite regular Borel measures on X and the space Mod(E,C(X)) of all continuous C(X)-module homomorphisms from E into C(X). Interpreting this result, we may say that the "external" dual of a bundle is obtained by tensoring the "internal" dual of the bundle with the "external" dual of the trivial bundle which has C(X) as its set of sections.

Having now revealed my original intentions, I should also say what I was able to achieve:

Firstly, I found it convenient to gather some known information from the literature for later references, and that is what is done in the first 10 sections. The informed reader will hardly find anything new here, an exception are perhaps the results concerning bundles of Ω-spaces, which present the common aspects of bundles of topological vector lattices, Banach algebras, C^*-algebras etc. Most of the other results here originated from papers of J.Dauns, K.H.Hofmann, L.Nachbin, S.Machado, J.B.Prolla, H.Möller, E.Behrends, E.M.Alfsen, E.G. Effros, A.Douady and L.Dal Soglio-Hérault and I apologize to all the others which are not mentioned here explicitely. To make these notes more self-contained, I included the proofs.

In section 11 we start with the development of a duality for bundles. Here the "dual unit ball" of the "unit ball" of a bundle of Banach spaces is introduced and it is shown how the upper semicontinuity (resp. continuity) of the norm of the bundle is reflected in this dual unit ball. Moreover, we discuss the relation between "stalkwise" convex subsets of the bundle and "stalkwise" convex subsets of the dual unit ball.

In sections 12, 13 and 14 we apply the results from section 11 to subbundles and quotients of bundles and discuss morphisms between bundles in general.

In paragraph 15 we take a closer look at the topology of the bundle space. Especially, we study the strength of separation in the bundle space and its relation to the closure of the "unit ball".

The theorem saying that every bundle with a Hausdorff bundle space whose stalks are of a fixed finite dimension n and whose base space is locally compact is in fact locally trivial stands in the center of sections 16 and 17.

In paragraph 18 we consider spaces of bounded linear operators with values in a space of sections in a bundle and prove a representation theorem for these spaces. An application of this representation theorem to spaces of compact operators yields a result concerning the approximation property of spaces of sections.

The study of the space of continuous C(X)-module homomorphisms into

C(X) is carried out in section 19. The main result presented here says that the space of sections in a "separable" bundle of Banach spaces with a compact base space and continuous norm admits "enough" continuous C(X)-module homomorphisms into C(X).

In section 20 we investigate to what extend the theorem of Mackey and Arens holds "internally" in the category of C(X)-modules

The last section is devoted to a treatment of the three problems mentioned above. The main part of this paragraph is taken from a joint work of Klaus Keimel and myself done in 1976 which was never publisched.

I am grateful to Klaus Keimel who always found the time for helpful conversations.

Notational remarks.

$\mathbb{K} = \mathbb{R}, \mathbb{C}$	is the field of real or complex numbers.
X	always denotes a topological space.
$C(X)$	stands for the algebra of all continuous $\mathbb{K}$-valued functions on X.
$C_b(X)$	denotes all continuous and bounded $\mathbb{K}$-valued functions on X.
conv M	is the convex hull of M.
$\overline{\text{conv}}\ M$	abbreviates the closed convex hull of M.
extr M	stands for the extreme points of M.

Compact and locally compact spaces are always understood to be Hausdorff, all $C_b(X)$-modules are unital and all topological vector spaces appearing in these notes are supposed to be locally convex.

1. Basic definitions

In many applications of functional analysis, the objects occuring there are not only topological vector spaces, but carry some extra structure turning them into algebras, vector lattices, C^*- algebras etc. As in these notes we would like to deal with all of them at the same time, we invent the following definition:

1.1 Definition. A *type* τ is a mapping $\tau : I \to \mathbb{N}$ from an index set I into the positive integers (including 0). A (*topological*) Ω-*space* of type τ is a pair (E,F), where E is a (topological) vector space and where $F = (f_i)_{i \in I}$ is a family of (continuous) mappings $f_i : E^{\tau(i)} \to E$.
An Ω-B-space is a topological Ω-space such that the underlying topological vector space is a Banach space.□

We shall often forget the type τ and the family F and speak simply of the Ω-space E.

1.2 Examples (i) Let $I = \{1\}$ and define $\tau(1) = 2$. Let (E,F) be an Ω-space of type τ. Then $F = (f_1)$ and $f_1 : E \times E \to E$ is a continuous mapping. Instead of $f_1(a,b)$ we shall write $a \cdot b$. If E happens to be a Banach space and if $\cdot$ satisfies the equations

$$||a \cdot b|| \leq ||a|| \, ||b||$$

$$a \cdot (b + c) = a \cdot b + a \cdot c$$

$$(a + b) \cdot c = a \cdot c + b \cdot c$$

$$(k \cdot a) \cdot b = k \cdot (a \cdot b) \qquad \text{for all } k \in \mathbb{K}$$

$$a \cdot (b \cdot c) = (a \cdot b) \cdot c$$

then E is a Banach algebra.

(ii) Let $\mathbb{K} = \mathbb{C}$ and let $I = \{1,2\}$. Define $\tau(1) = 2$ and $\tau(2) = 1$ and let (E,F) be an Ω-B-space of type τ. In this case we have $F = (f_1,f_2)$. The mapping $f_1 : E\times E \to E$ will again be written as multiplication and instead of $f_2(a)$ we shall write a^*.
If $\cdot$ and * satisfy the equations of example (i) and if in addition

$$(k\cdot a)^* = \bar{k}\cdot a^*$$
$$(a + b)^* = a^* + b^*$$
$$(a\cdot b)^* = b^*\cdot a^*$$
$$a^{**} = a$$
$$\|a^*\cdot a\| = \|a\|^2$$
$$\|a^*\| = \|a\|$$

then E is called a C^* - algebra.

(iii) If we let $\mathbb{K} = \mathbb{R}$, $I = \{1,2\}$ and $\tau(1) = \tau(2) = 2$, then we may define Banach lattices in a similar manner.

We now proceed with the central definition of the whole paper:

1.3 Definition. Let $p : E \to X$ be a mapping between two sets E and X. If $x \in X$ is an element of X, then the preimage $p^{-1}(x) =: E_x$ of x is called the *stalk* over x.
The *n-fold stalkwise product of* p is defined to be the set

$$\overset{n}{\vee} E = \{(a_1,\ldots,a_n) \in E^n : p(a_1) = \ldots = p(a_n)\}.$$

If there are mappings

$$\text{add} : E\vee E \to E$$
$$\text{scal} : \mathbb{K}\times E \to E$$
$$\mathbf{D} : X \to E$$

such that

$$p \circ \mathrm{add}(\alpha,\beta) = p(\alpha)$$
$$p \circ \mathrm{scal}(r,\alpha) = p(\alpha)$$
$$p \circ O(x) = x$$

and such that for every $x \in X$ the restrictions of add to $E_x \times E_x$ and scal to $\mathbb{K} \times E_x$ turn E_x into a vector space with respect to the operations $\alpha + \beta := \mathrm{add}(\alpha,\beta)$ and $r \cdot \alpha := \mathrm{scal}(r,\alpha)$ which has $O(x)$ as a zero, then the triple (E,p,X) is called a *fibred vector space*.

If $\tau : I \to \mathbb{N}$ is a type and if $F = (f_i)_{i \in I}$ is a family of mappings

$$f_i : \bigvee^{\tau(i)} E \to E$$

such that $p \circ f_i(\alpha_1,\ldots,\alpha_{\tau(i)}) = p(\alpha_1)$, then (E,p,X) is called a *fibred Ω-space*.

Now let $A \subset X$ be a subset of X. A *selection over* A is a mapping $\sigma : A \to E$ such that $p \circ \sigma = \mathrm{id}_A$. If σ_1 and σ_2 are two selections over A and if $k \in \mathbb{K}$ is a scalar, then we may define the sum $\sigma_1 + \sigma_2$ of σ_1 and σ_2 coordinatewise by

$$(\sigma_1 + \sigma_2)(a) := \sigma_1(a) + \sigma_2(a) \qquad \text{for all } a \in A.$$

Similary, the product $k \cdot \sigma$ is defined by

$$(k \cdot \sigma)(a) := k \cdot (\sigma(a)) \qquad \text{for all } a \in A.$$

With these operations, the selections over A form a vector space. If (E,p,X) is even a fibred Ω-space and if $\sigma_1,\ldots,\sigma_{\tau(i)}$ are selections over A, then $f_i(\sigma_1,\ldots,\sigma_{\tau(i)})$ defined by

$$f_i(\sigma_1,\ldots,\sigma_{\tau(i)})(a) := f_i(\sigma_1(a),\ldots,\sigma_{\tau(i)}(a)) \qquad \text{for all } a \in A$$

is also a selection over A. In this case, the set of all selections over A form an Ω-space. This Ω-space is exactly the product $\prod_{a\in A} E_a$ of the Ω-spaces E_a.

Let (E,p,X) be a fibred vector space. A mapping $\nu : E \to \mathbb{R}$ is called a *seminorm*, provided that for every $x \in X$ the mapping $\nu_{/E_x} : E_x \to \mathbb{R}$ is a seminorm on the vector space E_x in the usual sense.
A familiy of seminorms $(\nu_j)_{j\in J}$ is said to be *directed*, if for all pairs $j_0, j_1 \in J$ there is a $j \in J$ such that for all $\alpha \in E$ we have $\nu_{j_0}(\alpha),\ \nu_{j_1}(\alpha) \le \nu_j(\alpha)$.

If $(\nu_j)_{j\in J}$ is a family of seminorms on (E,p,X), then a selection σ over $A \subset X$ is called $(\nu_j)_{j\in J}$ - *bounded* (or just *bounded*, of no confusion about the family of seminorms in question is possible), if $\sup_{a\in A} \nu_j(\sigma(a))$ is finite for every $j \in J$. With $\Sigma_A(p)$ we denote the set of all bounded selections over A. If $X = A$, then we shall use the symbol $\Sigma(p)$ instead of $\Sigma_X(p)$. □

The following remark ist immediate:

1.4 Proposition. *For every* $A \subset X$ *the set* $\Sigma_A(p)$ *is a subspace of* $\prod_{a\in A} E_a$. *Moreover, for every* $j \in J$, *the mapping* $\Diamond_j : \Sigma_A(p) \to \mathbb{R} : \sigma \to \sup_{a\in A} \nu_j(\sigma(a))$ *is a seminorm on* $\Sigma_A(p)$. *If in addition the restriction of* ν_j *to the stalks* E_a, $a \in A$, *is a norm on* E_a, *then* $\Diamond_j$ *is a norm on* $\Sigma_A(p)$. □

If the sets E and X carry topologies and if $p : E \to X$ is continuous, then every continuous and bounded selection is called a *section*. With $\Gamma_A(p) \subset \Sigma_A(p)$ we denote the subset of all sections. As above, we write $\Gamma(p)$ instead of $\Gamma_X(p)$. If the domain of a section σ is open, then σ is called a *local section.*

1.5 Definition. Let τ be a type. A *bundle* of Ω-spaces of type τ is a fibred Ω-space (E,p,X) together with a directed famliy $(\nu_j)_{j\in J}$ of seminorms on E such that

I) E and X carry topologies and the mappings $p : E \to X$, add : $E \vee E \to E$, scal : $\mathbb{K}\times E \to E$, $0 : X \to E$ and $f_j : \bigvee^{\tau(j)} E \to E$ are continuous, where $\bigvee_n E$ carries the topology which is in-induced by the product topology on E^n.

II) The ε-*tubes*, i.e. the sets of the form

$$T(U,\sigma,\varepsilon,j) := \{\alpha \in E : p(\alpha) \in U \text{ and } \nu_j(\alpha - \sigma(p(\alpha)) < \varepsilon\}$$

where $U \subset X$ is open, $\sigma \in \Gamma_U(p)$, $\varepsilon > 0$ and $j \in J$, form a base for the topology on E.

III) For every choice of $j \in J$, every $\alpha \in E$ and every $\varepsilon > 0$ there is an open neighborhood U of $p(\alpha)$ and a section $\sigma \in \Gamma_U(p)$ such that $\nu_j(\alpha - \sigma(p(\alpha))) < \varepsilon$.

IV) We have $\alpha = 0(p(\alpha))$ if and only if $\nu_j(\alpha) = 0$ for all $j \in J$.

X is called the *base space* and E is called the *bundle space* [

1.6 Consequences. (i) *The mapping* p *is open.* (Indeed, we have $p(T(U,\sigma,\varepsilon,j)) = U$ and the sets of this form are a base for the topology by axiom II.)

(ii) *The seminorms* $\nu_j : E \to \mathbb{R}$ *are upper semicontinuous.* (This follows immediately from $\nu_j^{-1}(]-\infty,\varepsilon[) = T(X,0,\varepsilon,j)$.)

(iii) *For every* $j \in J$ *and every continuous selection* $\sigma : A \to E$ *the mapping* $x \to \nu_j(\sigma(x)) : A \to \mathbb{R}$ *is upper semicontinuous.*

(iv) *If* $A \subset X$ *is quasicompact, then every continuous selection* $\sigma : A \to E$ *is bounded.*

(v) *For every* $A \subset X$ *the set* $\Gamma_A(p)$ *is a linear subspace of* $\Sigma_A(p)$. *Moreover, if* A *is quasicompact, then* $\Gamma_A(p)$ *is an* Ω*-subspace of* $\prod_{a \in A} E_a$.
(Indeed, $\sigma_1 + \sigma_2 = \mathrm{add}(\sigma_1, \sigma_2)$ is continuous whenever σ_1 and σ_2 are continuous and $r \cdot \sigma = \mathrm{scal}(r, \sigma)$ is continuous whenever σ is. Whence $\Gamma_A(p)$ is a linear subspace of $\Sigma_A(p)$, as the triangle inequality yields the boundedness of $\sigma_1 + \sigma_2$.
Now suppose that $A \subset X$ is quasicompact and let $\sigma_1, \ldots, \sigma_{\tau(i)} \in \Gamma_A(p)$. Then $f_i(\sigma_1, \ldots, \sigma_{\tau(i)})$ is continuous by axiom I and therefore bounded by (iv), i.e. $f_i(\sigma_1, \ldots, \sigma_{\tau(i)})$ belongs to $\Gamma_A(p)$.)

(vi) *Let* $f \in C_b(X)$ *and let* $\sigma \in \Gamma_A(p)$. *If we define a selection* $f \cdot \sigma$ *by* $(f \cdot \sigma)(a) = f(a) \cdot \sigma(a)$ *for all* $a \in A$, *then we have* $f \cdot \sigma \in \Gamma_A(p)$. *Under this multiplication,* $\Gamma_A(p)$ *becomes a* $C_b(A)$ *- module.*

(vii) *For every* $\sigma \in \Gamma_U(p)$, *where* $U \subset X$ *is open, and for every* $x \in U$ *the family*

$$\{T(V, \sigma_{/V}, \varepsilon, j,) : x \in V \subset U,\ V \text{ open},\ \varepsilon > 0 \text{ and } j \in J\}$$

is a neighborhood base at $\sigma(x)$.
(To prove this assertion, let $\mathcal{O}$ ba any open set around $\sigma(x)$. By axiom II we may assume that $\mathcal{O} = T(W, \sigma', \varepsilon', j)$. Let $r := \nu_j(\sigma(x) - \sigma'(x)) < \varepsilon'$. By (iii) we can pick an open set V around x which is contained in W and satisfies $\nu_j(\sigma(y) - \sigma'(y)) < \frac{1}{2} \cdot (\varepsilon' + r)$ for all $y \in V$. If α is an element of E such that $p(\alpha) \in V$ and $\nu_j(\alpha - \sigma(p(\alpha))) < \frac{1}{2}(\varepsilon' - r)$, then from the triangle inequality we obtain $\nu_j(\alpha - \sigma'(p(\alpha))) < \varepsilon'$, i.e. $T(V, \sigma_{/V}, \frac{1}{2}(\varepsilon' - r), j) \subset \mathcal{O}$.)

(viii) *The stalks* E_x, $x \in X$, *equipped with the induced topologies, are locally convex topological vector spaces. Moreover, on* E_x *the topology induced by* E *and the topology generated by the sets* $\{\beta : p(\beta) = x \text{ and } \nu_j(\alpha - \beta) < \varepsilon\}$, *where* $\alpha \in E$, $j \in J$ *and* $\varepsilon > 0$, *agree.*

(Indeed, the topology induced by E is certainly coarser. Pick any $\alpha_o \in E$. Then, in the second topology, a typical open neighborhood of α_o looks like $\{\beta : p(\beta) = x \text{ and } \nu_j(\alpha_o - \beta) < \varepsilon\}$ for suitable $j \in J$ and $\varepsilon > 0$. By axiom III we can find an open neighborhood U of x and a local section $\sigma \in \Gamma_U(p)$ such that $\nu_j(\alpha_o - \sigma(x)) < \varepsilon/2$. Now the triangle inequality yields $\alpha_o \in E_x \cap T(U,\sigma,\varepsilon/2,j) \subset \{\beta : p(\beta) = x$ and $\nu_j(\alpha_o - \beta) < \varepsilon\}$.)

(ix) *The seminorms* $(\vartheta_j)_{j\in J}$ *on* $\Gamma_A(p)$ *defined by* $\vartheta_j(\sigma) = \sup_{a\in A} \nu_j(\sigma(a))$ *define a locally convex Hausdorff topology on* $\Gamma_A(p)$. *If* X *is quasicompact, then with respect to this topology,* $\Gamma(p)$ *is a topological* Ω*-space, i.e. the operations* $f_i : \Gamma(p)^{\tau(i)} \to \Gamma(p)$ *are continuous.*

(We have to show that for every $(\sigma_1,\ldots,\sigma_{\tau(i)}) \in \Gamma(p)^{\tau(i)}$, every $j_o \in J$ and every $\varepsilon > 0$ there is a $\delta > 0$ and a $j_1 \in J$ such that for every $\tau(i)$ - tuple $(\rho_1,\ldots,\rho_{\tau(i)}) \in \Gamma(p)^{\tau(i)}$ satisfying

$$\vartheta_{j_1}(\rho_1 - \sigma_1) < \delta,\ldots, \vartheta_{j_1}(\rho_{\tau(i)} - \sigma_{\tau(i)}) < \delta$$

the inequality

$$\vartheta_{j_o}(f_i(\rho_1,\ldots,\rho_{\tau(i)}) - f_i(\sigma_1,\ldots,\sigma_{\tau(i)})) < \varepsilon$$

holds.

Let $\sigma_1,\ldots,\sigma_{\tau(i)} \in \Gamma(p)$, $j_o \in J$ and $\varepsilon > 0$ be given. As the mapping $f_i : \bigvee^{\tau(i)} E \to E$ is continuous, the set $f_i^{-1}(T(X,f_i(\sigma_1,\ldots,\sigma_{\tau(i)}),\varepsilon/2, j_o))$ is open in $\bigvee^{\tau(i)} E$ and contains the $\tau(i)$ - tuple $(\sigma_1(x),\ldots, \sigma_{\tau(i)}(x))$ for every $x \in X$.

Fix $x \in X$ for a moment. As $\bigvee^{\tau(i)} E$ carries the topology induced by the product topology on E^n, we can find $\delta_{1,x},\ldots,\delta_{\tau(i),x} > 0$, elements $j_{1,x},\ldots,j_{\tau(i),x} \in J$ and open neighborhoods $U_{1,x},\ldots,U_{\tau(i),x}$ of x such that

$$\prod_{k=1}^{\tau(i)} T(U_{k,x},\sigma_k/U_{1,x},\delta_{k,x},j_{k,x}) \cap \bigvee^{\tau(i)} E \subset$$
$$\subset f_i^{-1}(T(X,f_i(\sigma_1,\ldots,\sigma_{\tau(i)}),\varepsilon/2,j_o))$$

(we have to use (vii) at this point!) Let $U_x = U_{1,x} \cap \ldots \cap U_{\tau(i),x}$, $\delta_x = \min\{\delta_{1,x},\ldots,\delta_{\tau(i),x}\}$ and choose $j_x \in J$ such that $\nu_{j_{1,x}}(\alpha),\ldots$ $\ldots,\nu_{j_{\tau(i),x}}(\alpha) < \nu_{j_x}(\alpha)$ for all $\alpha \in E$. Then we have

$$\prod_{k=1}^{\tau(i)} T(U_x,\sigma_k/U_x,\delta_x,j_x) \cap \bigvee^{\tau(i)} E \subset f_i^{-1}(T(X,f_i(\sigma_1,\ldots,\sigma_{\tau(i)}),\varepsilon/2,j_o)).$$

By construction, the open sets U_x, $x \in X$, cover X. As X is quasi-compact, there are $x_1,\ldots,x_n \in X$ such that $X = U_{x_1} \cup \ldots \cup U_{x_n}$. Define $\delta = \min\{\delta_{x_1},\ldots,\delta_{x_n}\}$ and choose $j_1 \in J$ such that $\nu_{j_{x_1}}(\alpha),\ldots,\nu_{j_{x_n}}(\alpha)$ $\le \nu_{j_1}(\alpha)$ for all $\alpha \in E$.

Now assume that the elements $\rho_1,\ldots,\rho_{\tau(i)} \in \Gamma(p)$ satisfy the inequalities $\vartheta_{j_1}(\rho_1 - \sigma_1),\ldots,\vartheta_{j_1}(\rho_{\tau(i)} - \sigma_{\tau(i)}) < \delta$. Let y be any element of X. Then there is a $k \in \{1,\ldots,n\}$ such that $y \in U_{x_k}$. This yields the following inequality:

$$\begin{aligned} \nu_{j_{x_k}}(\rho_l(y) - \sigma_l(y)) &\le \nu_{j_1}(\rho_l(y) - \sigma_l(y)) \\ &\le \vartheta_{j_1}(\rho_l - \sigma_l) \\ &< \delta \\ &\le \delta_{x_k} \qquad \text{for all } 1 \le k \le \tau(i), \end{aligned}$$

i.e. $(\rho_1(y),\ldots,\rho_{\tau(i)}(y)) \in f_i^{-1}(T(X,f_i(\sigma_1,\ldots,\sigma_{\tau(i)}),\varepsilon/2,j_1)$ or $\nu_{j_o}(f_i(\rho_1(y),\ldots,\rho_{\tau(i)}(y)) - f_i(\sigma_1(y),\ldots,\sigma_{\tau(i)}(y))) < \varepsilon/2$. To our satisfaction, this implies the inequality

$$\vartheta_{j_o}(f_i(\rho_1,\ldots,\rho_{\tau(i)}) - f_i(\sigma_1,\ldots,\sigma_{\tau(i)})) \le \varepsilon/2 < \varepsilon. \qquad)$$

(x) *If* $\sigma,\rho \in \Gamma_A(p)$ *and if* $\vartheta_j(\sigma) \le 1$, $\vartheta_j(\rho) \le 1$, *then* $\vartheta_j(f\cdot\sigma + (1 - f)\cdot\rho) \le 1$ *for all* $f \in C_b(A)$ *with* $0 \le f \le 1$. (This is an easy calculation using the definitions.)

(xi) *The mapping* add : $E \vee E \to E$ *is open.*

(Indeed, the map $\ell : (\alpha,\beta) \to (\alpha+\beta,\beta) : E\vee E \to E\vee E$ is a continuous bijection having the continuous inverse $\ell^{-1} : (\alpha,\beta) \to (\alpha-\beta,\beta) : E\vee E \to E\vee E$ and thus is a homeomorphismen. It is straightforward to see that the first projection $\pi : (\alpha,\beta) \to \alpha : E\vee E \to E$ is open. Whence the mapping add $= \pi\circ\ell$ is open, too.)□

1.7 Remarks. (i) If $A \subset X$ is not quasicompact, then $\Gamma_A(p)$ does not have to be an Ω-space, as the example (1.8(ii)) below shows. Hence the question, whether or not $\Gamma_A(p)$ is even a *topological* Ω-space in general makes only sense if the subset $A \subset X$ is quasicompact.
If every section $\sigma \in \Gamma_A(p)$ can be extended to a global section $\bar{\sigma} \in \Gamma(p)$ and if A is in fact quasicompact, then a closer look at the proof of (1.6(ix)) shows that $\Gamma_A(p)$ is indeed a topological Ω-space. The same remains true if X is quasicompact and $A \subset X$ is an arbitrary subset.
We shall return to the problem of extending sections in paragraph 4.

(ii) The property (1.6(x)) above was called C(X)-*convexity* by K.H.Hofmann in [Ho 75] and L.Nachbin in [Na 59].

1.8 Examples. (i) Let E be a topological Ω-space of a certain type $\tau : I \to \mathbb{N}$ with operations $(f_i)_{i\in I}$. Moreover, let $(\mu_j)_{j\in J}$ be an up-directed family of seminorms on E which generates the topology on E, i.e. the sets $\{\alpha \in E : \mu_j(\alpha) < \varepsilon\}$, where $\varepsilon > 0$ and $j \in J$, form a neighborhood base at $0 \in E$. Furthermore, let X be any topological space. We then can define a bundle of Ω-spaces in the following manner:
Let $\mathcal{E} := X\times E$ be equipped with the product topology and let $p : \mathcal{E} \to X$ be the first projection. Then, up to a natural homeomorphism, we have $\bigvee^n \mathcal{E} = X\times E^n$. If we define the operations add, scal, 0 and $(f_i)_{i\in I}$

"pointwise", i.e.

$$\begin{aligned}
\text{add} &: (x,\alpha,\beta) \to (x,\alpha + \beta) \;;\; E \vee E \to E \\
\text{scal} &: (r,(x,\alpha)) \quad (x,r\cdot\alpha) \;;\; E \vee E \to E \\
0 &: x \to (x,0) \qquad\qquad\quad ;\; E \vee E \to E \\
f_i &: (x,\alpha_1,\ldots,\alpha_{\tau(i)}) \to (x,f_i(\alpha_1,\ldots,\alpha_{\tau(i)})) \;;\; \bigvee^{\tau(i)} E \to E
\end{aligned}$$

then (E,p,X) is a fibred Ω-space.

Moreover, if we define seminorms $\nu_j : E \to \mathbb{R}$ on E by $\nu_j(x,\alpha) := \mu_j(\alpha)$, then (E,p,X) is a bundle of Ω-spaces, called the *trivial bundle with base space* X, *stalk* E *and seminorms* $(\nu_j)_{j\in J}$. To verify the axioms II and III, we have only to note that $\sigma \in \Gamma_A(p)$ is a section if and only if there is a bounded continuous map $\xi : A \to E$ such that $\sigma(x) = (x,\xi(x))$ for all $x \in A$, where $A \subset X$ is any subset.

Especially, we conclude that $\Gamma_A(p)$ is algebraically and topologically isomorphic to $C_b(A,E)$, where $C_b(A,E)$ denotes the topological vector space of all bounded E-valued continuous functions, equipped with the topology of uniform convergence on A.

(ii) Let E be a Banach space and let $f : E \to E$ be any continuous map which is not bounded. (Clearly, f cannot be linear!) Let $A \subset E$ be a bounded set such that $f(A)$ is not bounded. Then E together with the operation f is a topological Ω-space. We now form the trivial bundle with base space A and stalk E. As we just remarked, we have the isomorphy $\Gamma(p) \simeq C_b(A,E)$.

Moreover, if $\iota : A \to E$ denotes the inclusion map, then ι belongs to $C_b(A,E)$, but $f\circ\iota$ does not, as this mapping is not bounded. Whence $\Gamma(p)$ does not have to be an Ω-space in general.

(iii) However, let us remark that if E is a Banach algebra, a Banach lattice or a C^*-algebra, then $C_b(X,E)$ is of the same type for every topological space X.

(iv) Let $p : E \to X$ be a bundle of Ω-spaces and let $A \subset X$ be a subset. Moreover, let $E_A := p^{-1}(A)$ and let p_A be the restriction of p to E_A. Restricting the operations add, scal, 0 and $(f_i)_{i\in I}$ to $E_A \vee E_A$, $\mathbb{K}\times E_A$ and $\overset{\tau(i)}{\vee} E_A$ resp. we obtain a new bundle (E_A, p_A, A), called the *restriction of* (E,p,X) *to* A. It is clear that $\Gamma(p_A) = \Gamma_A(p)$.

(v) If M is a differentiable manifold and if $p : T \to M$ is the tangent bundle, then we also have a bundle in the sense of (1.5), if we take as a family of seminorms the usual Euclidean metric on the stalks. These bundles behave especially nice: They are locally trivial in the sense that every point x in the base space M has a neighborhood U such that the restriction (T_U, p_U, U) of (T,p,U) to U is isomorphic to a trivial bundle. We shall return to the precise definition and to a discussion of trivial bundles in a later paragraph.

Obviously, if $p : E \to X$ is a bundle with seminorms $(\nu_j)_{j\in J}$, then the seminorms $\hat{\nu}_j : \Sigma(p) \to \mathbb{R}$ induce also a locally convex Hausdorff topology on $\Sigma(p)$. Moreover, we have:

1.9 Proposition. *If* $p : E \to X$ *is a bundle, then* $\Gamma(p)$ *is closed in* $\Sigma(p)$.

Proof. Let σ be an element of $\Sigma(p)$ which belongs to the closure of $\Gamma(p)$. We have to show that σ is continuous.
Let us start with an element $x_o \in X$ and let O be an open neighborhood of $\sigma(x_o)$. By axiom 1.5.II we may assume that $O = \{\alpha \in E : : p(\alpha) \in U \text{ and } \nu_j(\alpha - \rho(p(\alpha))) < \varepsilon\}$, where $U \subset X$ is an open set around x_o, $\rho \in \Gamma_U(p)$ is a local section, $\varepsilon > 0$ and where ν_j is an appropriate seminorm. Let $\delta = \frac{1}{4}(\varepsilon - \nu_j(\sigma(x_o) - \rho(x_o)))$. As σ belongs to the closure of $\Gamma(p)$, we can find a section σ' such that $\hat{\nu}_j(\sigma - \sigma') < \delta$. Hence we have

$$\begin{aligned}\nu_j(\sigma'(x_o) - \rho(x_o)) &\le \nu_j(\sigma'(x_o) - \sigma(x_o)) + \nu_j(\sigma(x_o) - \rho(x_o))\\ &\le \vartheta_j(\sigma' - \sigma) + \varepsilon - 4\cdot\delta\\ &\le \delta + \varepsilon - 4\cdot\delta\\ &= \varepsilon - 3\cdot\delta\\ &< \varepsilon - 2\cdot\delta,\end{aligned}$$

i.e. $\sigma'(x_o) \in \{\alpha \in E : p(\alpha) \in U \text{ and } \nu_j(\alpha - \rho(p(\alpha))) < \varepsilon - 2\delta\}$. As the latter set is open and as σ' is continuous, we can find an open neighborhood $V \subset U$ of x_o such that $\nu_j(\sigma'(x) - \rho(x)) < \varepsilon - \delta$ for all $x \in V$. Whence for every $x \in V$ we have

$$\begin{aligned}\nu_j(\sigma(x) - \rho(x)) &\le \nu_j(\sigma'(x) - \rho(x)) + \nu_j(\sigma(x) - \sigma'(x))\\ &\le \nu_j(\sigma'(x) - \rho(x)) + \vartheta_j(\sigma' - \sigma)\\ &< \varepsilon - \delta + \delta = \varepsilon,\end{aligned}$$

i.e. $\sigma(V) \subset \mathcal{O}$. This shows the continuity of σ.□

Before we attack questions of completeness of $\Gamma(p)$, we need a little observation: Let $x \in X$. Then we have an evaluation map

$$\begin{aligned}\varepsilon_x : \Sigma(p) &\to E_x\\ \sigma &\to \sigma(x)\end{aligned}$$

and we shall denote the restriction of ε_x to $\Gamma(p)$ with the same symbol. As in both cases we have the inequality $\nu_j(\sigma(x)) \le \vartheta_j(\sigma)$, these mappings are continuous and linear. We shall see later, that in a large number of cases they are also quotient maps, i.e. they will be open and surjective.

1.10 Proposition. *Let* $p : E \to X$ *be a bundle. If all stalks* E_x, $x \in X$, *are complete (quasicomplete, semicomplete), then so are* $\Gamma(p)$ *and* $\Sigma(p)$.

Proof. As $\Gamma(p)$ is closed in $\Sigma(p)$, it is enough to prove these assertions for $\Sigma(p)$. Because in all three cases the proofs are analogous, we only give a proof for the case where all stalks are quasicomplete.

Thus, let $A \subset \Sigma(p)$ be a closed bounded subset and let $(\sigma_i)_{i\in I}$ be a Cauchy net in A. We have to show that the net $(\sigma_i)_{i\in I}$ converges. Firstly, let $A_x := \varepsilon_x(A)$. As ε_x is continuous, the set A_x is bounded in E_x and so is its closure. Now the quasicompletness of E_x yields that $\lim_{i\in I} \sigma_i(x)$ exists in E_x. Define a selection

$$\sigma : X \to E$$

by

$$x \to \lim_{i\in I} \sigma_i(x).$$

Then σ is bounded: Let ν_j be one of the seminorms on E. Then we find an $i_o \in I$ such that $\hat{\nu}_j(\sigma_{i_1} - \sigma_{i_2}) \leq 1$ for all $i_1, i_2 \geq i_o$. Using the triangle inequality and setting $i_2 = i_o$, we obtain $\nu_j(\sigma_{i_1}(x)) \leq \hat{\nu}_j(\sigma_{i_o}) + 1$ for all $i_1 \geq i_o$. Now let $x \in X$ be arbitrary. Then we have $\nu_j(\sigma_{i_1}(x)) \leq \hat{\nu}_j(\sigma_{i_o}) + 1$ and therefore $\nu_j(\sigma(x)) = = \nu_j(\lim_{i\in I} \sigma_i(x)) = \lim_{i\in I} \nu_j(\sigma_i(x)) \leq \hat{\nu}_j(\sigma_{i_o}) + 1$. As this holds for all $x \in X$, we conclude that $\hat{\nu}_j(\sigma) \leq \hat{\nu}_j(\sigma_{i_o}) + 1$ and whence σ is bounded.

Finally, the net $(\sigma_i)_{i\in I}$ converges to σ: Indeed, let $\varepsilon > 0$ and let ν_j be one of the seminorms. Then $\hat{\nu}_j(\sigma_{i_1} - \sigma_{i_2}) \leq \varepsilon/2$ for sufficently large $i_1, i_2 \in I$, i.e. $\nu_j(\sigma_{i_1}(x) - \sigma_{i_2}(x)) \leq \varepsilon/2$ for large i_1, i_2 and all $x \in X$. Sending i_2 to "infinity", we may conclude that $\nu_j(\sigma_{i_1}(x) - \sigma(x)) \leq \varepsilon/2$ for sufficently large i_1 and all $x \in X$, i.e. $\hat{\nu}_j(\sigma_{i_1} - \sigma) \leq \varepsilon/2 < \varepsilon$ for sufficently large i_1. This completes the proof. □

One of the most important classes of bundles are those for which every stalk is a Banach space. More precisely, we give the following definition:

1.11 Definition. Let τ be a type. A *Banach bundle of Ω-spaces* $p : E \to X$ is a bundle of Ω-spaces such that

(i) the family $(\nu_j)_{j \in J}$ of seminorms on E consists of one element $\|\cdot\|$ only (which then automatically induces a norm on each stalk).

(ii) every stalk E_x, $x \in X$, is a Banach space in the norm induced by $\|\cdot\|$. □

1.12 Proposition. *If* $p : E \to X$ *is a Banach bundle, then* $\Gamma(p)$ *is a Banach space in the norm given by* $\|\sigma\| := \sup_{x \in X} \|\sigma(x)\|$. □

2. Full bundles and bundles with completely regular base space

In the following two paragraphs, we try to simplify definition (1.5). This is possible, if we restrict ourselves to more special base spaces X.

Let us start with a definition:

2.1 Definition. A bundle $p : E \to X$ is called *full (locally full)*, if for every $\alpha \in E$ there is a (local) section σ such that $\alpha = \sigma(p(\alpha))$.
□

2.2 Proposition. *If X is completely regular, if $\sigma : U \to E$ is a local section and if $x \in U$, then there is a global section $\overline{\sigma} : X \to E$ such that $\overline{\sigma}(x) = \sigma(x)$.*

Proof. Let $U \subset X$ be open and let $\sigma : U \to E$ be a local section. As X is completely regular, we can find a neighborhood $V \subset U$ of x such that the closure $\overline{V}$ of V is still contained in U. Moreover, we can find a continuous mapping $f : X \to [0,1] \subset \mathbb{R}$ such that $f(x) = 1$ and such that f vanishes on the complement of V. We now define

$$\overline{\sigma} : X \to E$$

by

$$\overline{\sigma}(x) = \begin{cases} f(x)\cdot\sigma(x) & \text{if } x \in U \\ 0 & \text{if } x \in X \setminus \overline{V} \end{cases}$$

This definition makes sense, as on $U \cap (X \setminus \overline{V})$ we have $f(x) = 0$, i.e. $f(x)\cdot\sigma(x) = 0$. Clearly, $\overline{\sigma}$ is continuous, as it is continuous on the open sets U and $X \setminus \overline{V}$. □

2.3 Proposition. *If* X *is completely regular, then every locally full bundle over* X *is full.* □

If we restrict ourselves to locally full bundles, we need no longer the whole strenght of axiom (1.5.III). This means: If we make the third axiom in (1.5) stronger, then we may virtually weaken axiom II. To be precise:

2.4 Theorem. *Let* (E,p,X) *be a fibred* Ω*-space of type* τ *together with a directed family* $(\nu_j)_{j\in J}$ *of seminorms on* E *such that the following axioms are satisfied:*

I) E and X carry topologies and the mappings $p : E \to X$, $0 : X \to E$, add $: E \vee E \to E$, scal $: \mathbb{K}\times E \to E$ and $f_i : \overset{\tau(i)}{\bigvee} E \to E$ are continuous.

II) Given $x \in X$, then the sets of the form

$$\{\alpha \in E : p(\alpha) \in U,\ \nu_j(\alpha) < \varepsilon\}$$

where U is an open neighborhood of x, $\varepsilon > 0$ and $j \in J$, are an open neighborhood base at $0(x) \in E$.

III) Given $\alpha \in E$, then there is an open neighborhood U of $p(\alpha)$ and a local section $\sigma : U \to E$ such that $\sigma(p(\alpha)) = \alpha$.

IV) $\alpha = 0(p(\alpha))$ if and only if $\nu_j(\alpha) = 0$ for every $j \in J$.

Then (E,p,X) *is a bundle of* Ω*-spaces, which is, of course, locally full.*

Proof. We only have to check axiom (1.5.II)

Firstly, note that for every local section $\sigma : U \to E$ the mapping

$$T_\sigma : p^{-1}(U) \to p^{-1}(U)$$
$$\alpha \mapsto \alpha + \sigma(p(\alpha))$$

is continuous (being the composition of

$$\begin{array}{ccccccc} E_U & \to & E_U \check{\times} X & \to & E_U \vee E_U & \to & E_U \\ \alpha & \to & (\alpha, p(\alpha)) & \to & (\alpha, \sigma(p(\alpha)) & \to & \alpha + \sigma(p(\alpha)) \end{array}$$

where $E_U := p^{-1}(U)$ as in (1.8.iv)). Moreover, the mapping $T_{-\sigma}$ is a continuous inverse for T_σ . Thus, T_σ is a homeomorphismen.
Now fix $\alpha \in E$ and choose any local section $\sigma : U \to E$ such that $\sigma(p(\alpha)) = \alpha$. Then the homeomorphism T_σ transports the open neighborhood base $\{\{\beta \in E : \nu_j(\beta) < \varepsilon, p(\beta) \in V\} : \varepsilon > 0, j \in J, p(\alpha) \in V \subset U$, V open$\}$ of $O(p(\alpha))$ onto the open neighborhood base $\{\{\beta \in E : \nu_j(\beta - \sigma(p(\beta)) < \varepsilon$ and $p(\beta) \in V\}: p(\alpha) \in V \subset U$, V open, $\varepsilon > 0$ and $j \in J\}$ of $T_\sigma(O(p(\alpha))) = \alpha$.□

Now, of course, the question arises, which bundles are locally full. It turns out that under certain restrictions on the completeness of the stalks, all bundles with a locally regular base space are locally full and hence full by (2.3)

2.5 Proposition. *Let $p : E \to X$ be a bundle with seminorms $(\nu_j)_{j\in J}$ and with a completely regular base space X. If $x \in X$ is fixed and if $\varepsilon_x : \Gamma(p) \to E_x$ is the evaluation map, then for every $\alpha \in \varepsilon_x(\Gamma(p)) \subset E_x$ and every $j \in J$ we have* $\nu_j(\alpha) = \inf \{\hat{\nu}_j(\sigma) : \sigma(x) = \alpha, \sigma \in \Gamma(p)\}$.

Proof. Clearly, we always have the inequality

$$\nu_j(\alpha) \le \inf \{\hat{\nu}_j(\sigma) : \sigma(x) = \alpha, \sigma \in \Gamma(p)\}.$$

To verify the other inequality, pick an $\alpha \in E_x$, let $\varepsilon > 0$ and assume that there is a $\sigma' \in \Gamma(p)$ such that $\sigma'(x) = \alpha$. By (1.6.(iii)), the mapping $y \to \nu_j(\sigma'(y)) : X \to \mathbb{R}$ is upper semicontinuous. As $\nu_j(\sigma'(x)) = \nu_j(\alpha) < \nu_j(\alpha) + \varepsilon$, we may find an open neighborhood U of x such that $\nu_j(\sigma'(y)) < \nu_j(\alpha) + \varepsilon$ for all $y \in U$. Now the fact that X is completely regular yields a continuous function $f : X \to [0,1] \subset \mathbb{R}$ with

$f(x) = 1$ and $f(X \setminus U) = \{0\}$. Define $\sigma = f \cdot \sigma'$. Then for every $y \in U$ we have $\nu_j(\sigma(y)) = f(y) \cdot \nu_j(\sigma'(y)) \leq \nu_j(\alpha) + \varepsilon$ and for $y \in X \setminus U$ we may conclude that $\nu_j(\sigma(y)) = \nu_j(f(y) \cdot \sigma'(y)) = \nu_j(0 \cdot \sigma'(y)) = 0 < < \nu_j(\alpha) + \varepsilon$, whence $\hat{\nu}_j(\sigma) \leq \nu_j(\alpha) + \varepsilon$. As $\sigma(x) = \alpha$ and as $\varepsilon > 0$ was arbitrary, the proof is complete. □

2.6 Corollary. *If* $p : E \to X$ *is a bundle with a completely regular base space* X*, then the evaluation maps* $\varepsilon_x : \Gamma(p) \to E_x$, $x \in X$*, are open onto their images.* □

2.7 Corollary. *If* $p : E \to X$ *is a (locally) full bundle with a completely regular base space* X*, then for every* $x \in X$ *the evaluation map* $\varepsilon_x : \Gamma(p) \to E_x$ *is a quotient map.* □

2.8 Definition. Let (E,p,X) be a fibred vector space, let $(\nu_j)_{j \in J}$ be a family of seminorms on E and assume that the base space X carries a topology. If every point $x \in X$ has an open neighborhood U with the property that $\{j \in J : \nu_j(\alpha) \neq 0 \text{ for some } \alpha \in p^{-1}(U)\}$ is countable, then the family of seminorms $(\nu_j)_{j \in J}$ is called *locally countable.* □

If $p : E \to X$ is a bundle with a locally countable family $(\nu_j)_{j \in J}$ of seminorms, then every point $x \in X$ has a neighborhood U such that $\Gamma_U(p)$ is metrizable. Moreover, every stalk E_x is metrizable in the topology induced by E. This allows us to apply Banach's homomorphism theorem in the proof of the following result:

2.9 Theorem. *Let* $p : E \to X$ *be a bundle with a completely regular base space* X *and a locally countable family of seminorms* $(\nu_j)_{j \in J}$. *If all stalks* E_x, $x \in X$*, are complete, then* $p : E \to X$ *is a full bundle and the evaluation maps* $\varepsilon_x : \Gamma(p) \to E_x$ *are quotient maps.*

Proof. By (2.3) it is enough to show that $p : E \to X$ is locally full. Thus, let $x \in X$ and let U be an open neighborhood of x with the property that $\{j \in J : \nu_j(\alpha) \neq 0$ for some $\alpha \in p^{-1}(U)\}$ is countable. Then $\Gamma_U(p)$ is metrizable and, by (1.10), the vector space $\Gamma_U(p)$ is complete. Let $\varepsilon_x : \Gamma_U(p) \to E_x$ be the evaluation map. Then (1.5.III) and (2.2) applied to the bundle $p : p^{-1}(U) \to U$, show that the image of ε_x is dense in E_x. Moreover, by (2.6), the mapping ε_x is a topological homomorphism . Whence, by Banach's homomorphism theorem, the image of ε_x is closed. Thus, we may conclude that the mapping ε_x is surjective and therefore the bundle $p : E \to X$ is locally full. □

2.10 Corollary (Dupré). *If* X *is a completely regular topological space, then every bundle of Banach spaces* $p : E \to X$ *with base space* X *is full. Moreover, all the evaluation maps* $\varepsilon_x : \Gamma(p) \to E_x$ *are quotient maps of Banach spaces, i.e.* $\|\alpha\| = \inf \{ \|\sigma\| : \sigma \in \Gamma(p)$ and $\sigma(p(\alpha)) = \alpha\}$. □

We conclude this section with two results which we need for later references:

2.11 Proposition. *Let* $p : E \to X$ *be a bundle with a completely regular base space* X. *Then* $\{f\cdot\sigma : f \in C_b(X),\ f(x) = 0,\ \ \sigma \in \Gamma(p)\}$ *is dense in* $\{\sigma \in \Gamma(p) : \sigma(x) = 0\}$.

Proof. Let $\sigma \in \Gamma(p)$ be such that $\sigma(x) = 0$, let ν_j be one of the seminorms belonging to the bundle and suppose that $\varepsilon > 0$ is given. We shall complete the proof by constructing a continuous function $f \in C_b(X)$ with $f(x) = 0$ and $\hat{\nu}_j(f\cdot\sigma - \sigma) < \varepsilon$.
First of all, let U be an open neighborhood of x such that $\nu_j(\sigma(y)) < \varepsilon/2$ for all $y \in U$. Such an open set exists as $\nu_j(\sigma(x)) = 0$ and as

the mapping $y \to \nu_j(\sigma(y)$ is upper semicontinuous. As X is completely regular, there is a continuous mapping $f : X \to [0,1] \subset \mathbb{R}$ such that $f(x) = 0$ and $f(X \setminus U) = \{1\}$. An easy calculation shows that in fact we have $\hat{\nu}_j(f\cdot\sigma - \sigma) \leq \varepsilon/2 < \varepsilon$. □

2.12 Proposition. *Let* $p : E \to X$ *be a bundle with an arbitrary base space* X. *If all stalks are finite dimensional, then* $p : E \to X$ *is locally full.*

Proof. Let $x \in X$ and let $S_x := \{\sigma(x) : \sigma \in \Gamma_U(p)$ for some neighborhood U of $x\}$. By axiom (1.5.III), the set S_x is a dense subspace of the stalk E_x. As E_x is finite dimensional, we conclude that $S_x = E_x$.□

3. <u>Bundles with locally paracompact base spaces</u>

The definition of bundles given in (1.5) is rather complicated. The most annoying axiom is the postulate (1.5.II), because in many applications we would like to use bundles to describe topological vector spaces as spaces of sections in a bundle. Hence it is unsatisfactory to use sections already in the definition of bundles. But if the base space is locally paracompact, if the family of seminorms is locally countable and if the stalks are complete, the existence of "enough" local sections follows from the other axioms.

3.1 Definition. A topological space X is called *locally paracompact*, if every $x \in X$ has at least one closed and paracompact neighborhood. □

It can be shown that every locally paracompact space is completely regular. On the other hand, every locally compact space, every paracompact space and every locally metrisable space is locally paracompact. Moreover, in a locally paracompact space every point has a neighborhood base of closed and paracompact sets.

The central result of this section is stated as follows:

3.2 Theorem. *Let* (E,p,X) *be a fibered vector space, let* $(\nu_j)_{j\in J}$ *be a directed family of seminorms on* E *and assume that* E *and* X *carry topologies such that*

(O) p is open and continuous.

(I) the mappings add : $E \vee E \to E$ and scal : $\mathbb{K} \times E \to E$ are continuous.

(II) If $O_x \in E_x$ is the O-element of the stalk E_x, then the sets of the form $\{\alpha \in E : p(\alpha) \in U$ and $\nu_j(\alpha) < \varepsilon\}$ form an open neighborhood base at O_x, where U runs through all open neighborhoods of x, $\varepsilon > O$ and $j \in J$.

(III) $\alpha = O(p(\alpha))$ if and only if $\nu_j(\alpha) = O$ for all $j \in J$.

If all stalks are semicomplete in the topology induced by E and if X is locally paracompact and $|J| = 1$ __or__ if X is locally compact and if the family $(\nu_j)_{j\in J}$ is locally countable, then (E,p,X) is a full bundle.

This theorem has orginally been proved by A.Douady and L.Dal Soglio--Hêrault for bundles of Banach spaces (see the appendix of [Fe 77]). Our version here is, up to some corrections and modifications, due to H.Möller (see [Mö 78]).

We shall prove (3.2) in several steps. Firstly, we shall assume that the family $(\nu_j)_{j\in J}$ is countable and develop some results in this case. Hence we may assume that $J = \mathbb{N}$ and that $n \le m$ implies $\nu_n(\alpha) \le \nu_m(\alpha)$ for all $\alpha \in E$. Moreover, we shall always assume that X is completely regular.

3.3 Let us agree that we call a subset $U \subset E$ an ε-n-thin set, if for all $(\alpha,\beta) \in U{\times}U \cap (E{\vee}E)$ we have $\nu_n(\alpha - \beta) < \varepsilon$.
Fix an arbitrary $\alpha \in E$. If $n \in \mathbb{N}$ and if $\varepsilon > O$ are given, then α has an ε-n-thin neighborhood.
(Indeed, let $V = \{\alpha \in E : \nu_n(\alpha) < \varepsilon\}$. By axiom (3.2.II), the set $V \subset E$ is open. As the mapping $(\beta,\beta') \to \beta - \beta' : E{\vee}E \to E$ is continuous by (3.2.I) and as (α,α) is mapped onto $O(p(\alpha)) \in V$ under this map, there is a neighborhood U of α such that $U{\times}U \cap (E{\vee}E)$ is mapped into V. This set U then will be ε-n-thin.)

3.4 Let $\alpha \in E$, let Λ be a directed set and let $(\varepsilon_\lambda)_{\lambda\in\Lambda}$ be a net of strictly positive numbers such that $\lim_{\lambda\in\Lambda} \varepsilon_\lambda = 0$. Further, assume that for every $n \in \mathbb{N}$ there is an ε-n-thin neighborhood $U_{\varepsilon,n}$ of α such that

(i) $\lambda_1 \leq \lambda_2$ and $n_1 \leq n_2$ imply $U_{\lambda_2,n_2} \subset U_{\lambda_1,n_1}$.

(ii) The sets $\{p(U_{\lambda,n}) : \lambda \in \Lambda$ and $n \in \mathbb{N}\}$ form a neighborhood base of $p(\alpha)$.

Then the family $\{U_{\lambda,n} : \lambda \in \Lambda$ and $n \in \mathbb{N}\}$ is a neighborhood base of α.

(Put $x = p(\alpha)$ and $V_{\lambda,n} = p(U_{\lambda,n})$. Then it is easy to see that the sets $O_{\lambda,n} = \{\beta \in E : \nu_n(\beta) < \varepsilon_\lambda$ and $p(\beta) \in V_{\lambda,n}\}$, $\lambda \in \Lambda$ and $n \in \mathbb{N}$, are a neighborhood base at $O(x)$.

Now let W be any neighborhood of α. Since the mapping add is continuous and since $\text{add}(\alpha, O(x)) = \alpha$, we can find a neighborhood W' of α such that $\text{add}(W' \times O_{\lambda,n} \cap E \vee E) \subset W$, if only λ and n are large enough. As p is open by (3.2.0) and as $W' \cap U_{\lambda,n}$ is a neighborhood of α, we may choose $\lambda' \geq \lambda$ and $n' \geq n$ such that $V_{\lambda',n'} \subset p(W' \cap U_{\lambda,n})$.

We claim that $U_{\lambda',n'} \subset W$.

Firstly, we have $U_{\lambda',n'} \subset U_{\lambda',n'} \cap p^{-1}(p(W' \cap U_{\lambda,n}))$. Let $\beta \in U_{\lambda',n'}$. Then β belongs to $U_{\lambda,n}$. Moreover, there is an element $\beta' \in W' \cap U_{\lambda,n}$ such that $p(\beta) = p(\beta')$. Now $U_{\lambda,n}$ is ε_λ-n-thin, i.e. $\nu_n(\beta - \beta') < \varepsilon_\lambda$ and hence $\beta - \beta' \in O_{\lambda,n}$. This implies $\beta = \beta' + (\beta - \beta') =$
$= \text{add}(\beta, (\beta - \beta')) \in \text{add}(W' \times O_{\lambda,n} \cap E \vee E) \subset W$, i.e. $U_{\lambda',n'} \subset W$.)

3.5 let $\sigma : X \to E$ be a selection. We say that σ is ε-n-continuous, provided that

(i) The mapping $x \to \nu_n(\sigma(x)) : X \to \mathbb{R}$ is bounded.

(ii) For every $x \in X$ there is a neighborhood V of x and an ε-n-thin neighborhood U of $\sigma(x)$ such that $f(V) \subset U$.

3.6 If $\sigma : X \to E$ is a selection which is ε-n-continuous for every $\varepsilon > 0$ and every $n \in \mathbb{N}$, then σ is bounded and continuous.

Clearly, σ is bounded by (3.5(i)). Now fix $x \in X$. For every pair (m,n) of natural numbers let $U'_{m,n}$ be an $\frac{1}{m}$-n-thin neighborhood of $\sigma(x)$ such that $\sigma(p(U'_{m,n})) \subset U'_{m,n}$ (such an $U'_{m,n}$ exists by (3.5(ii)). Define

$$U_n = \cap \{U'_{l,k} : l,k \le n\}$$

Then U_n is $\frac{1}{n}$-n-thin, fulfilles still the inclusion $\sigma(p(U_n)) \subset U_n$, but we have in addition the relation $U_{n+1} \subset U_n$.

Next, let $(V_\lambda)_{\lambda\in\Lambda}$ be a decreasing neighborhood base at x and let $I = \mathbb{N}\times\Lambda$. For every $i = (n,\lambda)$ we define $\varepsilon_i = \frac{1}{n}$ and $U_{i,n} = p^{-1}(V_\lambda) \cap U_n$. Clearly, the net $(\varepsilon_i)_{i\in I}$ converges to 0, all the set $U_{i,n}$ are ε_i-n-thin and the sets of the form $p(U_{i,n})$ form a neighborhood base at x. Whence, by (3.4), the sets $(U_{i,n})_{i,n}$ form a neighborhood base at $\sigma(x)$. Moreover, by definition of the set $U_{i,n}$ we have $\sigma(p(U_{i,n})) \subset U_{i,n}$ for every $i \in I$ and every $n \in \mathbb{N}$, thus the map σ is continouos at x.)

3.7 Let $f_1,\dots,f_n : X \to \mathbb{K}$ be continuous $\mathbb{K}$-valued functions on X such that

$$\sum_{i=1}^{n} f_i(x) \neq 0 \qquad \text{for all } x \in X$$

Then the mapping

$$\Phi : \bigvee^{n} E \to E$$

$$(\alpha_1,\dots,\alpha_n) \to \sum_{i=1}^{n} f_i(p(\alpha_i))\cdot\alpha_i$$

is open and continuous.

(The continuity of Φ follows easily from the continuity of the mappings add and scal.

To show the openess of Φ, let $(\alpha_1,\dots,\alpha_n) \in \bigvee^{n} E$ and let $x = p(\alpha_1) = \dots = p(\alpha_n)$. We may assume without loss of generality that $f_1(x) \neq 0$. Let V be an open neighborhood of x such that $f_1(y) \neq 0$ for all $y \in V$.

As $\overset{n}{\vee} E \cap p^{-1}(V)^n$ is open in $\overset{n}{\vee} E$, it is enough to show that the restriction of Φ to $\overset{n}{\vee} E \cap p^{-1}(V)^n$ is open.

Define a mapping

$$\Psi : \overset{n}{\vee} E \cap p^{-1}(V)^n \to \overset{n}{\vee} E \cap p^{-1}(V)^n$$
$$(\alpha_1,\ldots,\alpha_n) \to (\sum_{i=1}^{n} f_i(p(\alpha_i))\cdot\alpha_i, \alpha_2,\ldots,\alpha_n).$$

Evidently, Ψ is continuous and has the continuous inverse

$$\Psi^{-1} : \overset{n}{\vee} E \cap p^{-1}(V)^n \to \overset{n}{\vee} E \cap p^{-1}(V)^n$$
$$(\alpha_1,\ldots,\alpha_n) \to (f_1(p(\alpha_1))^{-1}\cdot(\alpha_1 - \sum_{i=2}^{n} f_i(p(\alpha_i)))\cdot\alpha_i, \alpha_2,\ldots,\alpha_n)$$

and therefore is a homeomorphism.

By definition of the topology, the restriction of the first projection $\pi : (\alpha_1,\ldots,\alpha_n) \to \alpha_1 : \overset{n}{\vee} E \to E$ is open. As we have $\Phi = \pi\circ\Psi$, the mapping Φ is open as well.)

3.8 Let $f_1,\ldots,f_n : X \to \mathbb{R}$ be continuous real-valued functions, let $A_1,\ldots,A_n \subset E$ be ε-m-thin subsets of E and let

$$M := \sup \{ \sum_{i=1}^{n} |f_i(x)| : x \in p(A_1) \cap \ldots \cap p(A_n) \}.$$

If $\Phi : \overset{n}{\vee} E \to E$ is defined as in (3.7), then $\Phi(\overset{n}{\vee} E \cap A_1\times\ldots\times A_n)$ is $M\cdot\varepsilon$-m-thin.

(Indeed, let $\alpha,\beta \in \Phi(\overset{n}{\vee} E \cap A_1\times\ldots\times A_n)$ and let $x = p(\alpha) = p(\beta)$. Then we may find elements $\alpha_i,\beta_i \in A_i$, $1 \le i \le n$, such that $\alpha = \sum_{i=1}^{n} f_i(x)\cdot\alpha_i$ and $\beta = \sum_{i=1}^{n} f_i(x)\cdot\beta_i$. Since the sets A_i are ε-n-thin, we know that $\nu_n(\alpha_i - \beta_i) < \varepsilon$ for all $i \in \{1,\ldots,n\}$. This yields the inequality

$$\nu_n(\alpha - \beta) = \nu_n(\sum_{i=1}^{n} f_i(x)\cdot(\alpha_i - \beta_i))$$
$$\le \sum_{i=1}^{n} |f_i(x)|\cdot\nu_n(\alpha_i - \beta_i) < \sum_{i=1}^{n} |f_i(x)|\cdot\varepsilon \le M\cdot\varepsilon.)$$

From (3.7) and (3.8) it is easy to conclude

3.9 If $f_1,\dots,f_n : X \to [0,1] \subset \mathbb{R}$ are continuous functions such that $\sum_{i=1}^{n} f_i(x) = 1$ and if $\sigma_1,\dots,\sigma_n : X \to E$ are ε-n-continuous selections, then $\sum_{i=1}^{n} f_i \cdot \sigma_i$ is a ε-n-continuous selection, too.

3.10 Let $\varepsilon > 0$ and let $(\sigma_m)_{m\in\mathbb{N}} \subset \Sigma(p)$ be a sequence of ε-n-continuous selections and let $\sigma : X \to E$ be a selections such that $\lim_{m\to\infty} \vartheta_n(\sigma_m - \sigma) = 0$. Then σ is a $2\cdot\varepsilon$-n'-continuous selection for every $n' \leq n$.

(Firstly, by hypothesis there is a positive integer m such that

$$\vartheta_n(\sigma_m - \sigma) < \varepsilon/2$$

i.e.
$$\nu_n(\sigma_m(x) - \sigma(x)) < \varepsilon/2 \quad \text{for all } x \in X.$$

Fix $x_o \in X$ and let U be an open neighborhood of x_o and let V be an ε-n-thin neighborhood of $\sigma_m(x_o)$ such that $\sigma_m(U) \subset V$. Moreover, define

$$W := \{\alpha + \beta : p(\alpha) = p(\beta) \in U,\ \alpha \in V \text{ and } \nu_n(\beta) < \varepsilon/2\}$$

As the set $\{\alpha : \nu_n(\alpha) < \varepsilon/2 \text{ and } p(\alpha) \in U\}$ is open, the set W is open by (3.7). Moreover, $\{\alpha : \nu_n(\alpha) < \varepsilon/2 \text{ and } p(\alpha) \in U\}$ is ε-n-thin, thus W is $2\cdot\varepsilon$-n-thin by (3.8). Further, for every $x \in V$ we have $\sigma(x) = \sigma_m(x) + (\sigma(x) - \sigma_m(x)) \in W$, i.e. $\sigma(V) \subset W$. Finally, for all $n' \leq n$ and all $\alpha \in E$ we have $\nu_{n'}(\alpha) \leq \nu_n(\alpha)$. This implies that the set W is $2\cdot\varepsilon$-n-thin, too.

Now let $m \in \mathbb{N}$ be a natural number such that $\vartheta_n(\sigma_m - \sigma) \leq 1$. Then the triangle inequality yields for every $x \in X$ the relation

$$\begin{aligned}\nu_{n'}(\sigma(x)) &\leq \nu_n(\sigma(x))\\ &\leq \nu_n(\sigma_m(x) - \sigma(x)) + \nu_n(\sigma_m(x))\\ &\leq 1 + \nu_n(\sigma_m(x)).\end{aligned}$$

As $x \to \nu_n(\sigma_m(x))$ is bounded, so is $x \to \nu_{n'}(\sigma(x))$. Thus, the selection σ is $2\cdot\varepsilon$-n'-continuous.)

From (3.6) and (3.10) we may deduce:

3.11 Let $(\sigma_n)_{n\in\mathbb{N}}$ be a sequence of selections such that σ_n is ε_n-n-continuous. If the sequence $(\varepsilon_n)_{n\in\mathbb{N}}$ converges to 0 and if $\sigma : X \to E$ is a selection such that $\lim_{n\to\infty} \Phi_m(\sigma_n - \sigma) = 0$ for every $m \in \mathbb{N}$, then σ is a continuous and bounded section.

3.12 If $\varepsilon > 0$, if $n \in \mathbb{N}$ and if $\alpha \in E$ are given, then there is an ε-n-continuous selection $\sigma : X \to E$ such that $\sigma(p(\alpha)) = \alpha$. (Firstly, by (3.3) the element α has an open ε-n-thin neighborhood V'. Let $V := V' \cap \{\beta : \nu_n(\beta) < \nu_n(\alpha) + 1\}$. Then V is still an open ε-n-thin neighborhood of α. Furthermore, let $U = p(V)$ and let $x = p(\alpha)$. By the axiom of choice we can find a selection $\sigma' : U \to E$ such that $\sigma'(U) \subset V$ and $\sigma'(x) = \alpha$. Note that $x \to \nu_n(\sigma(x))$ is automatically bounded, as we have $\nu_n(\beta) < 1 + \nu_n(\alpha)$ for all $\beta \in V$. Next, choose a continuous function $f : X \to [0,1] \subset \mathbb{R}$ such that $f(x) = 1$ and $f^{-1}([0,1])^- \subset U$. We now define a selection σ by

$$\sigma(y) = \begin{cases} f(y)\cdot\sigma'(y) & \text{if } y \in U \\ 0 & \text{if } y \notin U \end{cases}$$

On U we have $\sigma = f\cdot\sigma' + (1 - f)\cdot 0$, whence σ is ε-n-continuous on U by (3.9). On the open set $V = X \setminus f^{-1}([0,1])^-$, the selection σ agrees with the continuous section 0. As U and V cover X and as $\sigma(x) = f(x)\cdot\sigma'(x) = 1\cdot\alpha = \alpha$, the proof is complete.)

3.13 Assume that X is compact <u>or</u> that $\nu_n = \nu_m$ for all $n,m \in \mathbb{N}$. Given an ε-n-continuous selection $\sigma : X \to E$ and a point $x_o \in X$, there is a $\varepsilon/2$-(n+1)- continuous selection $\sigma' : X \to E$ such that

(i) $\nu_n(\sigma' - \sigma) \leq \frac{3}{2}\cdot\varepsilon$

(ii) $\sigma'(x_o) = \sigma(x_o)$.

(Let $x \in X$. By (3.12) there is an $\varepsilon/2$-(n+1)-continuous selection σ_x such that $\sigma_x(x) = \sigma(x)$. Obviously, σ_x is also $\varepsilon/2$-n-continuous. Choose an open ε-n-thin neighborhood V of $\sigma(x)$, an $\varepsilon/2$-n-thin neighborhood of $\sigma_x(x)$ $(= \sigma(x))$ and an open neighborhood U of x such that $\sigma(U) \subset V$ and $\sigma_x(U) \subset W$. As $V \cap W$ is an open set around $\sigma(x)$ and as p is open, the set $U_x := U \cap p(V \cap W)$ is an open neighborhood of x. Moreover, if y is an element of U_x, then there is an $\alpha \in V \cap W$ such that $y = p(\alpha)$. Hence we obtain the inequality

$$\begin{aligned}\nu_n(\sigma_x(y) - \sigma(y)) &\leq \nu_n(\sigma_x(y) - \alpha) + \nu_n(\alpha - \sigma(y)) \\ &\leq \varepsilon + \varepsilon/2 \\ &= \frac{3}{2}\cdot\varepsilon\end{aligned}$$

as $\alpha, \sigma(y) \in V$ and $\alpha, \sigma_x(y) \in W$.

Let U be any open neighborhood of x_o such that $\overline{U} \subset U_{x_o}$. Replacing each U_x, $x_o \neq x \in X$, by $U_x \setminus \overline{U}$, we obtain the following:

There is an open cover $(U_i)_{i\in I}$ of X and $\varepsilon/2$-(n+1)-continuous selections $(\sigma_i)_{i\in I}$ such that

(i) x_o belongs to exactly one U_{i_o} and for this index i_o we have $\sigma_{i_o}(x_o) = \sigma(x_o)$.

(ii) For every $i \in I$ and every $x \in U_i$ we have the inequality $\nu_n(\sigma_i(x) - \sigma(x)) \leq \frac{3}{2}\cdot\varepsilon$.

As X is paracompact, we may find a partition of unity $(f_i)_{i\in I}$ subordinate to the cover $(U_i)_{i\in I}$. By the property (i) above we may conclude that $f_{i_o}(x_o) = 1$ and $f_i(x_o) = 0$ for $i \neq i_o$. We now define our selection σ' by

$$\sigma' = \sum_{i\in I} f_i\cdot\sigma_i$$

Note that this sum is locally finite, i.e. each point has a neighborhood U such that $\{i : f_i(U) \neq \{0\}\}$ is finite. Whence (3.9), applied to these neighborhoods shows that the restriction $\sigma'_{/U} : U \to E_U := p^{-1}(U)$ of σ' to U is $\varepsilon/2$-(n+1)-continuous. Especially, σ' will satisfy the property (3.5(ii)), which is a local property.

Before we show the n+1-boundedness of σ', we prove that σ' has the properties (i) and (ii) of (3.13):

As $\sigma'(x_o) = \sum_{i\in I} f_i(x_o)\cdot\sigma_i(x_o) = f_{i_o}(x_o)\cdot\sigma_{i_o}(x_o) = \sigma(x_o)$, the property (ii) is satisfied.

Next, we check property (i): Let $x \in X$ be fixed and let $i \in I$ be any element. Then we have either $x \notin U_i$, in which case we conclude $f_i(x)\cdot\nu_n(\sigma_i(x) - \sigma(x)) = 0$; or we have $x \in U_i$, and then it is true that $f_i(x)\cdot\nu_n(\sigma_i(x) - \sigma(x)) \leq \frac{3}{2}\cdot f_i(x)\cdot\varepsilon$. Whence for all $i \in I$ and all $x \in X$ we have $f_i(x)\cdot\nu_n(\sigma_i(x) - \sigma(x)) \leq \frac{3}{2} f_i(x)\cdot\varepsilon$. As the $f_i(x)$ sum up to 1, this yields for every $x \in X$ the inequality

$$\begin{aligned}
\nu_n(\sigma'(x) - \sigma(x)) &= \nu_n\Big(\sum_{i\in I} f_i(x)\cdot\sigma_i(x) - \Big(\sum_{i\in I} f_i(x)\Big)\cdot\sigma(x)\Big) \\
&= \nu_n\Big(\sum_{i\in I} f_i(x)\cdot(\sigma_i(x) - \sigma(x))\Big) \\
&\leq \sum_{i\in I} f_i(x)\cdot\nu_n(\sigma_i(x) - \sigma(x)) \\
&\leq \sum_{i\in I} f_i(x)\cdot \frac{3}{2}\cdot\varepsilon = \frac{3}{2}\cdot\varepsilon
\end{aligned}$$

and therefore (i) holds.

Finally, we have to show the n+1-boundedness of σ':

Firstly, assume that X is compact. As the family $(f_i)_{i\in I}$ is locally finite, an easy compactness argument shows that there is a finite subset $J \subset I$ such that $f_i \neq 0$ if and only if $i \in J$. Therefore the sum we used in the definition of σ' is actually finite. Thus, σ' is $\varepsilon/2$-n+1-continuous by (3.9)

Now assume that $\nu_n = \nu_m$ for all $n,m \in \mathbb{N}$. Then the triangle inequality and the fact that $\vartheta_{n+1}(\sigma' - \sigma) = \vartheta_n(\sigma' - \sigma)$ yield that the map $x \to \nu_n(\sigma'(x))$ is bounded, i.e. the property (i) of (3.5) holds. As we checked the property (ii) of (3.5) already, the proof is complete.)

Applying (3.12) and (3.13), we obtain after an obvious recursion:

3.14 Assume that either X is compact or that $\nu_n = \nu_m$ for every pair of natural numbers n,m. If $\alpha \in E$ is given, then there exists a sequence $(\sigma_n)_{n\in\mathbb{N}}$ of selections such that

(i) σ_n is $(\frac{1}{2})^n$-n-continuous.

(ii) $\vartheta_n(\sigma_{n+1} - \sigma_n) \le \frac{3}{2}\cdot(\frac{1}{2})^n$ for all $n \in \mathbb{N}$

(iii) $\sigma_n(p(\alpha)) = \alpha$ for all $n \in \mathbb{N}$

3.15 If $(\sigma_n)_{n\in\mathbb{N}}$ is a sequence of selection which satisfies the properties (i), (ii) and (iii) of (3.14), then there is a selection σ such that $\lim_{n\to\infty} \vartheta_m(\sigma_n - \sigma) = 0$ for all $m \in \mathbb{N}$. This selection satisfies in addition the equation $\sigma(p(\alpha)) = \alpha$.

(Let $\varepsilon > 0$ and let $n_o \in \mathbb{N}$. Choose a natural number N such that $n_o \le N$ and such that

$$\sum_{n=N}^{\infty} \frac{3}{2} (\frac{1}{2})^n < \varepsilon.$$

Then for all natural numbers $m \ge n \ge N$ we have the inequality

$$\begin{aligned} \vartheta_{n_o}(\sigma_m - \sigma_n) &= \vartheta_{n_o}(\sum_{i=0}^{m-n-1} (\sigma_{n+i+1} - \sigma_{n+i})) \\ &\le \sum_{i=0}^{m-n-1} \vartheta_{n_o}(\sigma_{n+i+1} - \sigma_{n+i}) \\ &\le \sum_{i=0}^{m-n-1} \vartheta_{n+i}(\sigma_{n+i+1} - \sigma_{n+i}) \\ &\le \sum_{i=0}^{\infty} (\frac{3}{2})\cdot(\frac{1}{2})^{n+i} \le \sum_{i=N}^{\infty} (\frac{3}{2})\cdot(\frac{1}{2})^{i} < \varepsilon. \end{aligned}$$

In particular, since for every $x \in X$ the sets $\{\alpha \in E : \nu_n(\alpha) < \varepsilon\}$, $n \in \mathbb{N}$, and $\varepsilon > 0$, form a neighborhood base of 0 in the stalk E_x with the induced topology, the sequence $(\sigma_n(x))_{n\in\mathbb{N}}$ is a Cauchy sequence. As all the stalks are assumed to be semicomplete, $\lim_{n\to\infty} \sigma_n(x)$ exists in E_x. Now define $\sigma : X \to E$ by $\sigma(x) := \lim_{n\to\infty} \sigma_n(x)$. It remains to show that $\lim_{n\to\infty} \vartheta_{n_o}(\sigma_n - \sigma) = 0$ for all $n_o \in \mathbb{N}$.
Thus, let $\varepsilon > 0$ and let $n_o \in \mathbb{N}$. As we have seen above, we can find a natural number N such that $\vartheta_{n_o}(\sigma_m - \sigma_n) < \varepsilon/2$ for all $m,n \geq N$. Whence, for every $x \in X$ and all $m,n \geq N$ we have $\nu_{n_o}(\sigma_m(x) - \sigma_n(x)) <$ $< \varepsilon/2$. Sending n to infinity, this yields $\nu_{n_o}(\sigma_m(x) - \sigma(x)) \leq \varepsilon/2$ for all $m \geq N$. Since this inequality holds for all $x \in X$ and since $\varepsilon > 0$ was arbitrary, this yields $\lim_{n\to\infty} \vartheta_{n_o}(\sigma_n - \sigma) = 0$.
Clearly, by the definition of σ, we have $\sigma(p(\alpha)) = \lim_{n\to\infty} \nu_n(p(\alpha)) =$ $= \lim_{n\to\infty} \alpha = \alpha$.)

Now (3.11), (3.14) and (3.15) allow us to conclude:

3.16 If X is compact <u>or</u> if $\nu_n = \nu_m$ for all $n,m \in \mathbb{N}$, then for every $\alpha \in E$ there is a continuous section $\sigma : X \to E$ such that $\sigma(p(\alpha)) = \alpha$.

It is now easy to finish the proof of our theorem: Let $\alpha \in E$ and let $x = p(\alpha)$. Choose a paracompact (resp. compact) neighborhood U of x (such that only countably many of the seminorms have value different form 0 on $p^{-1}(U)$). Now (3.16) applied to $(p^{-1}(U),p,U)$ yields a bounded and continuous section $\sigma : U \to E$ such that $\sigma(x) = \alpha$. Now (2.3) and (2.4) together with the fact that every locally paracompact space is completely regular show that (E,p,X) is a full bundle.

4. Stone - Weierstraß theorems for bundles

The classical theorem of Stone and Weierstraß has been generalized in many ways (see [Bu 58], [Br 59], [Bi 61], [Gl 63], [We 65], [NMP 71], [Ho 75], [Gi 77], [Mö 78]). The results which will be represented in this section are due to Machado, Nachbin and Prolla ([NMP 71]) and K.H.Hofmann ([Ho 75]).

4.1 Definition. Let $p : E \to X$ be a bundle. A family $(f_i)_{i \in I} \subset C(X)$ is called *locally finite*, if every point $x \in X$ has a neighborhood U such that $\{i : f_i(y) \neq 0$ for some $y \in U\}$ is finite.
A subspace $F \subset \Gamma(p)$ is called *fully additive*, if for every locally finite family $(f_i)_{i \in I} \subset C_b(X)$ and every family $(\sigma_i)_{i \in I} \subset F$ the selection $\sum_{i \in I} f_i \cdot \sigma_i$ belongs to F, provided that this selection is bounded.
A subspace $F \subset \Gamma(p)$ is called *stalkwise dense*, if for each $x \in X$ the set $\varepsilon_x(F)$ is dense in the stalk E_x, where $\varepsilon_x : \Gamma(p) \to E_x$ denotes the evaluation map. □

It is obvious that every fully additive subspace F of $\Gamma(p)$ is also a $C_b(X)$-submodule.

We now turn to our Stone-Weierstraß theorem, which in this form is due to Hofmann, Machado, Nachbin and Prolla:

4.2 Theorem. *Let* $p : E \to X$ *be a bundle and let* $F \subset \Gamma(p)$ *be a fully additive and stalkwise dense subspace of* $\Gamma(p)$. *Then under each of the following two conditions,* F *is dense in* $\Gamma(p)$:

(i) The base space X is compact.

(ii) The base space X is paracompact and $p : E \to X$ is a bundle of normed spaces, i.e. $|J| = 1$.

Proof. Let $\sigma \in \Gamma(p)$. Then we have to show that for every $\varepsilon > 0$ and every seminorm ν_j belonging to the bundle there is a section $\rho \in F$ with $\hat{\nu}_j(\sigma - \rho) < \varepsilon$.
Firstly, fix an arbitrary point $x \in X$. As F is stalkwise dense, there is a section $\rho_x \in F$ such that $\nu_j(\rho_x(x) - \sigma(x)) < \varepsilon/2$. By (1.6.(iii)) we can find an open neighborhood U_x of x such that $\nu_j(\rho_x(y) - \sigma(y)) < \varepsilon/2$ for all $y \in U_x$.
Now the open sets U_x, $x \in X$, cover X. As X is at least paracompact, we may choose a partition of unity $(f_x)_{x \in X}$ subordinate to the open cover $(U_x)_{x \in X}$. Especially, the family $(f_x)_{x \in X} \subset C_b(X)$ is locally finite.
We define

$$\rho := \sum_{x \in X} f_x \cdot \rho_x$$

Then $\rho : X \to E$ is a continuous selection. Moreover, ρ is a bounded selection: Indeed, if X is compact, then the boundedness of ρ follows from (1.6.(iv)). On the other hand, if $p : X \to E$ is a bundle of normed spaces, then the family of seminorms $(\nu_j)_{j \in J}$ consists of one element only, which is just the seminorm ν_j we used above. in this case we only have to show that the mapping $x \to \nu_j(\rho(x)) : X \to \mathbb{R}$ is bounded. As we shall see in a moment, we have $\nu_j(\sigma(y) - \rho(y)) < \varepsilon$ for all $y \in X$. Thus, in this case the boundedness of ρ follows from the triangle inequality and the boundedness of σ.
Thus, in both cases the selection ρ will belong to F. It remains to show that $\hat{\nu}_j(\sigma - \rho) < \varepsilon$.
Let us start with a $y \in Y$. Then we may compute:

$$\nu_j(\sigma(y) - \rho(y)) = \nu_j(\,1\cdot\sigma(y) - \sum_{x\in X} f_x(y)\cdot\rho_x(y))$$

$$= \nu_j(\sum_{x\in X} f_x(y)\cdot\sigma(y) - \sum_{x\in X} f_x(y)\cdot\rho_x(y))$$

$$= \nu_j(\sum_{x\in X} f_x(y)\cdot(\sigma(y) - \rho_x(y)))$$

$$\le \sum_{x\in X} f_x(y)\cdot\nu_j(\sigma(y) - \rho_x(y)).$$

Now we have either $y \in U_x$ and hence $\nu_j(\sigma(y) - \rho_x(y)) < \varepsilon/2$, or we have $y \notin U_x$, in which case $f_x(y) = 0$. Thus, in both cases we may conclude that $f_x(y)\cdot\nu_j(\sigma(y) - \rho_x(y)) \le f_x(y)\cdot\varepsilon/2$. This implies the inequality $\nu_j(\sigma(y) - \rho(y)) \le \frac{1}{2}\cdot\varepsilon\cdot\sum f_x(y) = \varepsilon/2$, i.e. $\hat{\nu}_j(\sigma - \rho) \le \varepsilon/2 < \varepsilon$. □

For convenience we state the version of (4.2) which we shall use most often:

4.3 Corollary. *Let* $p : E \to X$ *be a bundle of Banach spaces over a compact base space* X *and let* $F \subset \Gamma(p)$ *be a stalkwise dense* $C(X)$*-submodule of* $\Gamma(p)$. *Then* F *is dense in* $\Gamma(p)$. □

We conclude this section with an application of our Stone-Weierstraß theorem. In (2.9) we have seen that for completely regular base spaces and "locally" completely metrizable $\Gamma(p)$ the evaluation maps $\varepsilon_x : \Gamma(p) \to E_x$ are quotient maps.
Now suppose that $A \subset X$ is any subset. Then we also have an evaluation map $\varepsilon_A : \sigma \to \sigma_{/A} : \Gamma(p) \to \Gamma_A(p)$. Again we ask for conditions under which this map is a quotient map. The Stone-Weierstraß helps to find an answer:

4.4 Theorem. *Let* $p : E \to X$ *be a bundle with a countable family of seminorms and assume that* X *is normal and that all stalks are complete. If* $A \subset X$ *is compact, then every section* $\sigma : A \to E$ *may be extended to a global section. Moreover, the evaluation map* $\varepsilon_A : \Gamma(p) \to \Gamma_A(p)$ *is a quotient map.*

Proof. As $\Gamma(p)$ and $\Gamma_A(p)$ are complete and metrizable, using Banach's homomorphism theorem it is enough to show that ε_A is a topological homomorphism, i.e. ε_A is open onto its image, and that the image of ε_A is dense in $\Gamma_A(p)$.
The fact that ε_A is topological follows as in (2.5) and (2.6) using the normality of X instead of the regularity in the proofs.
It remains to show that the image of ε_A is dense in $\Gamma_A(p)$:
As $p : E \to X$ is a full bundle by (2.9), the image of ε_A is stalk-wise dense. Moreover, the image of ε_A is a $C(A)$-submodule of $\Gamma_A(p)$:
Let $\sigma \in \Gamma_A(p)$ be of the form $\sigma = \varepsilon_A(\sigma')$ for a certain $\sigma' \in \Gamma(p)$ and let $f \in C(A)$ be a $\mathbb{K}$-valued continuous function on A. As X is normal, we can find an extension $f' \in C_b(X)$ of f. Now we have $f \cdot \sigma =$
$= f'_{/A} \cdot \sigma'_{/A} = (f' \cdot \sigma')_{/A} = \varepsilon_A(f' \cdot \sigma')$.
Now an application of the Stone-Weierstraß theorem (4.2) completes the proof. □

The following corollary, which is analougos to (2.10), has been proved and reproved by several authors : K.H.Hofmann credited this result to M.Dupré, J.M.G.Fell has shown it in [Fe 77], and J.W. Kitchen and D.A.Robbins proved an even stronger version for compact base spaces in [KR 80]: Every section $\sigma : A \to E$ may be extended *under the preservation of norm* to a global section, provided that X is compact, $A \subset X$ is closed and $p : E \to X$ is a bundle of Banach spaces.

4.5 Corollary. *If* $p : E \to X$ *is a bundle of Banach spaces over a normal base space* X *and if* $A \subset X$ *is compact, then every section* $\sigma : X \;\; \to E$ *may be extended to a global section. Moreover, the evaluation map* $\varepsilon_A : \Gamma(p) \to \Gamma_A(p)$ *is a quotient map of Banach spaces.* □

5. An alternative description of spaces of sections: Function modules

There is an alternative way of describing spaces of section which does not make use of the topology on the bundle space $\mathcal{E}$. For bundles of Banach spaces, this description is due to F.Cunningham (see [Cu 67]), and for the general setting we refer to the paper of Nachbin, Machado and Prolla (see [NMP 71]).

Suppose that we begin with a bundle $p : \mathcal{E} \to X$ of Ω-spaces of a certain type $\tau : I \to \mathbb{N}$ and suppose that the base space X is compact. Let $E := \Gamma(p)$. Then we know from (1.6), (1.9) and (2.2) that E has the following properties:

(FM1) For every $x \in X$ there is a topological vector space E_x; the topology of E_x is induced by a family of seminorms $(\nu_j^x)_{j \in J}$.

(FM2) E is a closed linear subspace of $\prod_{x \in X}^{\infty} E_x$, where

$$\prod_{x \in X}^{\infty} E_x = \{\sigma \in \prod_{x \in X} E_x : \sup_{x \in X} \nu_j^x(\sigma(x)) < \infty \text{ for all } j \in J\},$$

equipped with the topology induced by the seminorms $(\hat{\nu}_j)_{j \in J}$ given by

$$\hat{\nu}_j(\sigma) = \sup_{x \in X} \nu_j^x(\sigma(x)).$$

(FM3) The set $\{\sigma(x) : \sigma \in E\}$ is dense in E_x for every $x \in X$.

(FM4) The mapping $x \to \nu_j^x(\sigma(x)) : X \to \mathbb{R}$ is upper semicontinuous for every $\sigma \in E$ and every $j \in J$.

(FM5) E is a $C_b(X)$-module relative to the multiplication given by $(f \cdot \sigma)(x) := f(x) \cdot \sigma(x)$ for all $x \in X$, $f \in C_b(X)$ and $\sigma \in E$.

(FM6) Each of the E_x is a topological Ω-space of type τ and E is a topological Ω-subspace of $\prod^{\infty}_{x\in X} E_x$, i.e. if $i \in I$ and if f_i is one of the additional operations, then for all $\sigma_1,\ldots,\sigma_{\tau(i)} \in E$ we have $f_i(\sigma_1,\ldots,\sigma_{\tau(i)}) \in E$ and the mapping

$$f_i : E^{\tau(i)} \to E$$

is continuous, where $f_i(\sigma_1,\ldots,\sigma_{\tau(i)})(x) = f_i(\sigma_1(x),\ldots,\sigma_{\tau(i)}(x))$.

5.1. Definition. Let X be a topological space and let $\tau : I \to \mathbb{N}$ be a type. If E is a topological vector space satisfying (FM1) - (FM5), then E is called a *function module with seminorms* $(\nu_j^x)_{j\in J}$. Moreover, if the axiom (FM6) holds, then E is called an *Ω-function module of type* τ. The space X is called the *base space*; the vector spaces E_x, $x \in X$, are called the *stalks* of the function module. □

We shall see that every Ω-function module is in fact (isomorphic to) the space of all sections of a bundle of Ω-spaces of the same type, provided that the base space is compact.

Let us start with the so called "standard construction of bundles", which is due to K.H.Hofmann (see [Ho 75]) in the case of Banach bundles and has been generalized by H.Möller ([Mö 78]) to our present situation:

5.2 Let $(E_x)_{x\in X}$ be a family of vector spaces. If we set $\mathcal{E} := \bigcup_{x\in X} \{x\}\times E_x$ and $p : (x,\alpha) \to x : \mathcal{E} \to X$, then $(\mathcal{E},p,X)$ is a fibred vector space. As we already remarked in (1.4), every element of the cartesian product $\prod_{x\in X} E_x$ may be viewed as a selection of $(\mathcal{E},p,X)$. Further, let $(\nu_j^x)_{j\in J}$ be a directed family of seminorms on E_x generating a Hausdorff topology. Then we may define a directed family

$(\nu_j)_{j\in J}$ of seminorms on $\mathcal{E}$ by $\nu_j((x,\alpha)) = \nu_j^x(\alpha)$.

Suppose now that X carries a topology and that E is a linear subspace of $\prod_{x\in X} E_x$ such that (FM3) and (FM4) are satisfied. Then we have

5.3 The sets of the form $T(U,\sigma,\varepsilon,j) := \{\alpha \in \mathcal{E} : p(\alpha) \in U$ and $\nu_j(\alpha - \sigma(p(\alpha))) < \varepsilon\}$, where $U \subset X$ is open, $\sigma \in E$, $\varepsilon > 0$ and $j \in J$, form a base for a topology on $\mathcal{E}$.

(We have to show that for each $\alpha \in T(U_1,\sigma_1,\varepsilon_1,j_1) \cap T(U_2,\sigma_2,\varepsilon_2,j_2)$ there are an open set $U_3 \subset X$, an element $\sigma_3 \in E$, an $\varepsilon_3 > 0$ and an $j_3 \in J$ such that $\alpha \in T(U_3,\sigma_3,\varepsilon_3,j_3) \subset T(U_1,\sigma_1,\sigma_1,j_1) \cap T(U_2,\sigma_2,\varepsilon_2,j_2)$. Thus, let us suppose that such an α is given. Let $j_3 \in J$ be an index such that $j_1,j_2 \le j_3$. Moreover, we define ε_3 by the formula

$$\varepsilon_3 := \frac{1}{2} \min \{\varepsilon_k - \nu_{j_k}(\alpha - \sigma_k(p(\alpha))) : k = 1,2\}$$

Further, use (FM3) to find an element $\sigma_3 \in E$ such that

$$\nu_{j_3}(\alpha - \sigma_3(p(\alpha))) < \varepsilon_3.$$

Then for k = 1,2 we have

$$\begin{aligned}
\nu_{j_k}(\sigma_k(p(\alpha)) - \sigma_3(p(\alpha))) &\le \nu_{j_k}(\sigma_k(p(\alpha)) - \alpha) + \nu_{j_k}(\alpha - \sigma_3(p(\alpha))) \\
&\le \nu_{j_k}(\sigma_k(p(\alpha)) - \alpha) + \nu_{j_3}(\alpha - \sigma_3(p(\alpha))) \\
&< \nu_{j_k}(\sigma_k(p(\alpha)) - \alpha) + (2\cdot\varepsilon_3 - \varepsilon_3) \\
&\le \nu_{j_k}(\sigma_k(p(\alpha)) - \alpha) + (\varepsilon_k - \\
&\qquad - \nu_{j_k}(\alpha - \sigma_k(p(\alpha))) - \varepsilon_3) \\
&= \varepsilon_k - \varepsilon_3
\end{aligned}$$

By (FM 4) we now can pick an open neighborhood $U_3 \subset U_1 \cap U_2$ of $p(\alpha)$ such that $\nu_{j_k}(\sigma_3(y) - \sigma_k(y)) < \varepsilon_k - \varepsilon_3$ for all $y \in U_3$. We claim that $T(U_3,\sigma_3,\varepsilon_3,j_3) \subset T(U_1,\sigma_1,\varepsilon_1,j_1) \cap T(U_2,\sigma_2,\varepsilon_2,j_2)$: Indeed, let $\beta \in T(U_3,\sigma_3,\varepsilon_3,j_3)$. Then $p(\beta) \in U_3 \subset U_k$, k=1,2, and therefore

$$\nu_{j_k}(\beta - \sigma_k(p(\beta))) \le \nu_{j_k}(\beta - \sigma_3(p(\beta))) + \nu_{j_k}(\sigma_3(p(\beta)) - \sigma_k(p(\beta)))$$
$$< \nu_{j_3}(\beta - \sigma_3(p(\beta))) + (\varepsilon_k - \varepsilon_3)$$
$$< \varepsilon_3 + (\varepsilon_k - \varepsilon_3) = \varepsilon_k.$$

From now on, the set E always carries this topology.

5.4 If $\sigma \in E$, then the mapping $\sigma : X \to E$ is continuous. (Let $x_o \in X$ and let $\sigma(x_o) \in T(\sigma',U,\varepsilon,j)$. Then $\nu_j((\sigma - \sigma')(x_o)) < \varepsilon$. By (FM4) there is an open neighborhood $V \subset U$ of x_o such that $\nu_j(\sigma(y) - \sigma'(y)) < \varepsilon$ for all $y \in V$. Clearly, this implies $\sigma(V) \subset \subset T(U,\sigma',\varepsilon,j)$, i.e. σ is continuous at x_o.)

5.5 The mappings add : $E \vee E \to E$ and scal : $\mathbb{K}\times E \to E$ are continuous. (Let $(\alpha,\beta) \in E\vee E$ and let $T(U,\sigma,\varepsilon,j)$ be a neighborhood of $\alpha + \beta$. Let $\delta = \frac{1}{4}(\varepsilon - \nu_j(\alpha + \beta - \sigma(p(\alpha))))$. Then there are elements $\sigma_1,\sigma_2 \in E$ such that $\nu_j(\sigma_1(p(\alpha)) - \alpha) < \delta$ and $\nu_j(\sigma_2(p(\beta)) - \beta) < \delta$. For these elements we have

$$\nu_j(\sigma(p(\alpha)) - (\sigma_1(p(\alpha)) + \sigma_2(p(\beta))) \le$$
$$\le \nu_j(\sigma(p(\alpha)) - (\alpha + \beta)) + \nu_j(\alpha - \sigma_1(p(\alpha))) + \nu_j(\beta - \sigma_2(p(\beta)))$$
$$< \varepsilon - 2\cdot\delta.$$

Pick an open neighborhood $V \subset U$ of $p(\alpha)$ such that

$$\nu_j(\sigma(y) - \sigma_1(y) - \sigma_2(y)) < \varepsilon - 2\cdot\delta \quad \text{for all } y \in V.$$

Then the set $T(V,\sigma_1,\delta,j)\times T(V,\sigma_2,\delta,j) \cap E\vee E$ is a neighborhood of (α,β) and for all $(\alpha',\beta') \in T(V,\sigma_1,\delta,j)\times T(V,\sigma_2,\delta,j) \cap E\vee E$ we obtain

$$\nu_j(\alpha' + \beta' - \sigma(p(\alpha'))) \le \nu_j(\alpha' - \sigma_1(p(\alpha'))) + \nu_j(\beta' - \sigma_2(p(\beta'))) +$$
$$+ \nu_j((\sigma_1 + \sigma_2 - \sigma)(p(\alpha')))$$
$$< \delta + \delta + \varepsilon - 2\cdot\delta$$
$$= \varepsilon$$

This shows the continuity of add.

Now let $(r_o, \alpha_o) \in \mathbb{K} \times E$ and let $T(U, \sigma, \varepsilon, j)$ be an open neighborhood of $r_o \cdot \alpha_o$. In this case we choose $\delta = \varepsilon - \nu_j(\sigma(p(\alpha_o)) - r_o \cdot \alpha_o)$. Pick an element $\sigma' \in E$ such that $r_o \cdot \nu_j(\sigma'(p(\alpha_o)) - \alpha_o) < \delta/2$. Then we obtain

$$\begin{aligned}\nu_j(r_o \cdot \sigma'(p(\alpha_o)) - \sigma(p(\alpha_o))) &\le \nu_j(r_o \cdot \sigma'(p(\alpha_o)) - r_o \cdot \alpha_o) + \\ &\qquad + \nu_j(r_o \cdot \alpha_o - \sigma(p(\alpha_o))) \\ &< \varepsilon - \delta/2.\end{aligned}$$

Hence there is an open neighborhood $V \subset U$ of $p(\alpha_o)$ such that

$$\nu_j(r_o \cdot \sigma'(y) - \sigma(y)) < \varepsilon - \delta/2 \quad \text{for all } y \in V.$$

Choose a real number $0 < \varepsilon'$ such that

$$\varepsilon' \cdot (\varepsilon' + \hat{\nu}_j(\sigma')) + |r_o| \cdot \varepsilon' + (\varepsilon - \delta/2) < \varepsilon.$$

If $|r - r_o| < \varepsilon'$ and if $\alpha \in T(V, \sigma', \varepsilon', j)$, then we calculate:

$$\begin{aligned}\nu_j(r \cdot \alpha - \sigma(p(\alpha))) &\le \nu_j((r - r_o) \cdot \alpha) + \nu_j(r_o \cdot \alpha - \sigma(p(\alpha))) \\ &\le |r - r_o| \cdot \nu_j(\alpha) + \nu_j(r_o \cdot (\alpha - \sigma'(p(\alpha)))) + \\ &\qquad + \nu_j(r_o \cdot \sigma'(p(\alpha)) - \sigma(p(\alpha))) \\ &\le |r - r_o| \cdot (\nu_j(\alpha - \sigma'(p(\alpha))) + \nu_j(\sigma'(p(\alpha)))) + \\ &\qquad + |r_o| \cdot \varepsilon' + (\varepsilon - \delta/2) \\ &\le \varepsilon' \cdot (\varepsilon' + \nu_j(\sigma'(p(\alpha)))) + |r_o| \cdot \varepsilon' + (\varepsilon - \delta/2) \\ &< \varepsilon.\end{aligned}$$

This implies that the mapping scal is continuous, too.)

5.6 The mappings $p : E \to X$ and $0 : X \to E$ are continuous.
(This follows immediatly from the definitions of the topology on E and (5.4).)

5.7 If $U \subset X$ is an open set, if $\sigma : U \to E$ is continuous, if $\varepsilon > 0$

and if $j \in J$, then the set $\{\alpha \in E : p(\alpha) \in U$ and $\nu_j(\alpha - \sigma(p(\alpha))) < \varepsilon\}$ is open in E.

(Indeed, the mapping $T_\sigma : \alpha \to \alpha + \sigma(p(\alpha)) : p^{-1}(U) \to p^{-1}(U)$ is continuous, as add, σ and p are continuous. Because T_σ has continuous inverse $T_{-\sigma}$, it is a homeomorphism. Now note the set in question is the image under T_σ of the open set $\{\alpha \in E : p(\alpha) \in U$ and $\nu_j(\alpha) < \varepsilon\}$ and therefore is relatively open in $p^{-1}(U)$. As $p^{-1}(U)$ is open itself, the result follows.)

5.8 **Proposition.** *Let* $(E_x)_{x\in X}$ *be a family of topological vector spaces whose resp. topologies are induced by seminorms* $\nu_j^x : E_x \to \mathbb{R}$ $j \in J$. *Let* E *be a subspace of* $\prod_{x\in X}^{\infty} E_x$ *and assume that the index set* X *carries a topology such that* (FM3) *and* (FM4) *are satisfied. Then there is a bundle* $p : E \to X$ *with stalks (isomorphic to)* E_x, $x \in X$, *such that* E *is (up to isomorphy) a subspace of* $\Gamma(p)$. □

5.9 **Theorem.** *If* X *is a compact topological space, then there is a one-to-one correspondence between the class of all bundles with base space* X *and the class of all function modules with base space* X. *More explicitly:*

If $p : E \to X$ *is a bundle, then* $\Gamma(p)$ *is a function module with stalks* $p^{-1}(x)$, $x \in X$.

Conversely, if E *is a function module with base base* X *and stalks* E_x, $x \in X$, *then the construction given in* (5.2) *yields a bundle* $p_E : E_E \to X$ *and these two operations are inverse to each other. Especially, if* E *is a function module with a compact base space* X, *then there is a bundle* $p_E : E_E \to X$ *having the same stalks as* E *such that* $E \simeq \Gamma(p)$ *and this isomorphism preserves the* C(X)*-module structure.*

Proof. Let us start with a function module E with base space X. Then the construction given in (5.2) yields a bundle $p_E : \mathcal{E}_E \to X$ such that E may be viewed as a closed subspace of $\Gamma(p_E)$. From (FM3) and (FM5) we know that E is stalkwise dense and a C(X)-submodule of $\Gamma(p_E)$. Now the Stone-Weierstraß theorem (4.2) shows that E is dense in $\Gamma(p_E)$. As E was already closed in $\Gamma(p_E)$, we obtain $E \simeq \Gamma(p_E)$.

Conversely, let us suppose that we are given a bundle $p : \mathcal{E} \to X$ and let $E := \Gamma(p)$. Clearly, as we remarked at the beginning of this section, E is a function module. It is obvious that we may identify the sets $\mathcal{E}$ and $\mathcal{E}_E$ and the projections p and p_E. We only have to show that this identification is a homeomorphism for the topologies on $\mathcal{E}$ and $\mathcal{E}_E$ resp. Firstly, note that the topology on $\mathcal{E}_E$ is certainly coarser than the topology on $\mathcal{E}$, as we used only global sections in the definition of the topology on $\mathcal{E}_E$. Whence, applying (5.7), it remains to prove that every local section of $p : \mathcal{E} \to X$ is continuous when viewed as a selection of $p_E : \mathcal{E}_E \to X$.
Thus, let $\sigma : U \to \mathcal{E}$ be a local section and let $x_o \in U$. We want to show that the mapping $\sigma : U \to \mathcal{E}_E$ is continuous at x_o. Pick neighborhoods V,W of x_o such that $\overline{V} \subset W \subseteq \overline{W} \subset U$ and let $f : X \to [0,1]$ be a continuous function such that $f(\overline{V}) = \{1\}$ and $f(X \setminus W) = \{0\}$. Define a global section $\overline{\sigma} : X \to \mathcal{E}$ by $\overline{\sigma}(x) = 0$ for $x \in X \setminus \overline{W}$ and $\overline{\sigma}(x) = f(x)\cdot\sigma(x)$ if $x \in U$. (This definition makes sense , as on $(X \setminus \overline{W}) \cap U = U \setminus \overline{W}$ we have $f(x)\cdot\sigma(x) = 0\cdot\sigma(x) = 0$.) Then $\overline{\sigma}$ is continuous as it is continuous on the open sets $X \setminus \overline{W}$ and U and as these open sets cover X. Because $p : \mathcal{E} \to X$ and $p_E : \mathcal{E}_E \to X$ have the same global sections by the part of the theorem already verified, $\overline{\sigma} : X \to \mathcal{E}_E$ is continuous.
To finish the proof, we only have to remark that $\overline{\sigma}$ and σ agree on the open neighborhood V of x_o. □

We now turn our attention to Ω-function modules:

Let $\tau : I \to \mathbb{N}$ be a type and let E be an Ω-function module with stalks $(E_x)_{x \in X}$, base space X and seminorms $(\nu_j^x)_{j \in J}$. If (E,p,X) is the bundle constructed in (5.2), then for every $i \in I$ we may define a function

$$f_i \quad : \quad \bigvee^{\tau(i)} E \quad \to \quad E$$
$$f_i((x,\alpha_1),\ldots(x,\alpha_{\tau(i)})) := (x,f_i(\alpha_1,\ldots,\alpha_{\tau(i)}))$$

Of course, we hope that we obtain a bundle of Ω-spaces in this manner. I do not know an answer to this question in general, but I can offer some partial solutions:

Firstly, we take a closer look at the proof of the continuity of the mapping add : $E \vee E \to E$ in (5.5). Then we will recognize that the key inequality looks as follows:

$$\nu_j(\mathrm{add}(\alpha_1,\alpha_2) - \mathrm{add}(\beta_1,\beta_2)) \leq \nu_j(\alpha_1 - \beta_1) + \nu_j(\alpha_2 - \beta_2).$$

This means that the continuity of the addition in topological vector spaces is in some sense uniform for all vector spaces. This does not have to be true for the additional operations $(f_i)_{i \in I}$ a priori. Whence, if we would attempt to modify the proof of the continuity of add to show the continuity of the $(f_i)_{i \in I}$, we would have to postulate something like an "uniform continuity" for the $(f_i)_{i \in I}$. In this case, we would obtain some very technical condition like

(*) For every $i \in I$, every $j \in J$ and every $\varepsilon > 0$ there is an $j' \in J$ and a $\delta > 0$ such that for all $x \in X$ and all elements $\alpha_1,\beta_1,\ldots,\alpha_{\tau(i)},\beta_{\tau(i)} \in E_x$ the inequalities $\nu_{j'}(\alpha_1 - \beta_1),\ldots$ $\ldots,\nu_{j'}(\alpha_{\tau(i)} - \beta_{\tau(i)}) < \delta$ imply $\nu_j(f_i(\alpha_1,\ldots,\alpha_{\tau(i)}) -$ $- f_i(\beta_1,\ldots,\beta_{\tau(i)})) < \varepsilon$.

Let us agree that we call an Ω-space E an *uniform Ω-function module* provided that (*) holds.

A straightforward modification of the proof of the continuity of the mapping add now shows:

5.10 Proposition. *If* E *is a uniform* Ω*-function module with base space* X, *then there is a bundle of* Ω*-spaces* $p : E \to X$ *such that* E *is (topologically and algebraically isomorphic to) a closed subspace of* $\Gamma(p)$. □

There are certain cases for which the uniform continuity of the additional operations follows automatically. For instance, every function module of Banach lattices is uniform. This follows from the inequality $\|a \vee b - c \vee d\| \leq \|a - c\| + \|b - d\|$.
On the other hand, if we restrict the class of base spaces, then the uniformity is not needed:

5.11 Proposition. *If* X *is a completely regular topological space and if* E *is an* Ω*-function module with base space* X *satisfying the stronger axiom*

(FFM3) For every $x \in X$ we have $E_x = \{\sigma(x) : \sigma \in E\}$.

then there is a bundle of Ω*-spaces* $p : E \to X$ *such that* E *is (topologically and algebraically isomorphic to) a closed subspace of* $\Gamma(p)$.

Proof. Let $p : E \to X$ be the bundle constructed in (5.2),(5.3). By (5.8) it remains to show that $p : E \to X$ is a bundle of Ω-spaces. Firstly, note that we may use the proof of (2.5) to obtain from (FFM3) the condition

(**) For every $x \in X$ and every $\alpha \in E_x$ we have $\nu_j^x(\alpha) = \inf \{\hat{\nu}_j(\sigma) : \sigma(x) = \alpha, \sigma \in E\}$.

Clearly, condition (**) and (FFM3) imply

(**) If $\sigma \in E$, $\alpha \in E_x$ and if $\nu_j^x(\alpha - \sigma(x)) < \varepsilon$, then there is a $\sigma' \in E$ with $\nu_j(\sigma - \sigma') < \varepsilon$ and $\sigma'(x) = \alpha$.

We now want to show that the mappings $f_i : \bigvee^{\tau(i)} E \to E$, $i \in I$, are continuous.

Let $i \in I$, let $(\alpha_1,\ldots,\alpha_{\tau(i)}) \in \bigvee^{\tau(i)} E$ and let $\mathcal{O}$ be an open neighborhood of $f_i(\alpha_1,\ldots,\alpha_{\tau(i)})$. Further, let $x_o := p(\alpha_1) = \ldots = p(\alpha_{\tau(i)})$. By (FFM3) and (1.6(vii)) we may assume that there is an element $\sigma \in E$, an open neighborhood U of x_o, an $\varepsilon > 0$ and an $j \in J$ such that $\mathcal{O} = T(U,\sigma,\varepsilon,j)$. Applying (FFM3) once again, we find elements $\sigma_1,\ldots$ $\ldots,\sigma_{\tau(i)} \in E$ such that $\sigma_k(x_o) = \alpha_k$ for all $1 \leq k \leq \tau(i)$. As $f_i(\sigma_1,\ldots,\sigma_{\tau(i)})(x_o) = f_i(\sigma_1(x_o),\ldots,\sigma_{\tau(i)}(x_o)) = f_i(\alpha_1,\ldots,\alpha_{\tau(i)}) \in \mathcal{O}$, there is an open neighborhood $U' \subset U$ of x_o, an $\varepsilon' > 0$ and an $j' \in J$ such that $T(U',f_i(\sigma_1,\ldots,\sigma_{\tau(i)}),\varepsilon',j') \subset \mathcal{O}$ (use (1.6(viii))).

We now apply the fact that E itself is a topological Ω-space, i.e. that the operations $f_i : E^{\tau(i)} \to E$ are continuous. Whence it is possible to choose $\delta > 0$ and $j'' \in J$ such that the inequalities

$$\vartheta_{j''}(\sigma_1 - \sigma_1'),\ldots, \vartheta_{j''}(\sigma_{\tau(i)} - \sigma_{\tau(i)}') < \delta$$

imply

$$\vartheta_{j'}(f_i(\sigma_1,\ldots,\sigma_{\tau(i)}) - f_i(\sigma_1',\ldots,\sigma_{\tau(i)}')) < \varepsilon'.$$

We show that $f_i : \bigvee^{\tau(i)} E \to E$ maps the open neighborhood

$$T(U',\sigma_1,\delta,j'')\times\ldots\times T(U',\sigma_{\tau(i)},\delta,j'') \cap \bigvee^{\tau(i)} E$$

of $(\alpha_1,\ldots,\alpha_{\tau(i)})$ into $\mathcal{O}$.

Indeed, let $(\beta_1,\ldots,\beta_{\tau(i)})$ belong to the first of these sets and let $x := p(\beta_1) = \ldots = p(\beta_{\tau(i)})$. By (**) there are elements $\sigma_1',\ldots,\sigma_{\tau(i)}'$ $\in E$ with $\beta_k = \sigma_k'(x)$ and $\vartheta_{j''}(\sigma_k - \sigma_k') < \delta$ for all $1 \leq k \leq \tau(i)$.

Whence we may conclude that

$$\nu_{j'}(f_i(\sigma_1,\ldots,\sigma_{\tau(i)}) - f_i(\sigma'_1,\ldots,\sigma'_{\tau(i)})) < \varepsilon'$$

and especially

$$\nu_{j'}(f_i(\sigma_1(x),\ldots,\sigma_{\tau(i)}(x)) - f_i(\beta_1,\ldots,\beta_{\tau(i)})) < \varepsilon'.$$

This gives us finally the relation

$$f_i(\beta_1,\ldots,\beta_{\tau(i)}) \in T(U',f_i(\sigma_1,\ldots,\sigma_{\tau(i)}),\varepsilon',j') \subset O. \quad \square$$

5.12 Remark. Let us point out that theorem (2.9) also holds for function modules, as in its proof we only used the properties listed in (FM1) - (FM5). This means that every function module with a completely regular base space, complete stalks and a locally countable family of seminorms satisfies (FFM3).

The following theorem is analogous to (5.9):

5.13 Theorem. *Let* X *be a compact topological space and let* $\tau : I \to \mathbb{N}$ *be a type. If* $p : E \to X$ *is a bundle of* Ω*-spaces of type* τ*, then* $\Gamma(p)$ *is an* Ω*-function module of type* τ.
Conversely, if E *is an* Ω*-function module of type* τ *with base space* X*, which satisfies the stronger axiom* (FFM3)*, then there is a bundle* $p_E : E_E \to X$ *of* Ω*-spaces such that* E *is (topologically and algebraically isomorphic to) the* Ω*-space* $\Gamma(p)$.
Moreover, these two operations set up a one-to-one correspondence between the class of all Ω*-function modules satisfying* (FFM3) *and the class of all full bundles.* $\square$

We conclude the section with a few examples:

5.14 Bundles over the circle. Let $S^1 = \{z \in \mathbb{C} : |z| = 1\}$ be the unit circle. Furthermore, let F be any Banach space with norm $\|\cdot\|$ and let $T : F \to F$ be a linear contraction. Moreover, define

$$E_T := \{\sigma \in C([0,2\pi],F) : \sigma(2\pi) = T(\sigma(0))\}.$$

Clearly, E_T is a Banach space under pointwise addition und scalar multiplication when equipped with the norm $\|\cdot\|$ given by

$$\|\sigma\| := \sup \{ \|\sigma(x)\| : 0 \le x \le 2\pi\}.$$

Moreover, E_T may be viewed as a function module over S^1: For every $z \in S^1$ let $E_z := F$, equipped with the norm $\|\cdot\|$. Then we may identify E with a closed subspace of $\prod^{\infty}_{z\in S^1} E_z$ by sending $\sigma \in E_T$ to $\hat{\sigma}$, where for $0 \le \phi < 2\pi$ we define

$$\hat{\sigma}(e^{i\phi}) := \sigma(\phi).$$

If we do so, E_T becomes a function module with base space S^1, stalks $E_z = F$, and (semi-)norm $\|\cdot\|$. The only axiom which requires a little bit of work in verifying is the axiom (FM4):
Let $\sigma \in E_T$. We have to show that the mapping $z \to \|\hat{\sigma}(z)\| : S^1 \to \mathbb{R}$ is upper semicontinuous. The only problematic point of S^1 is $z = 1$. Thus, let us assume that $\|\hat{\sigma}(1)\| < \varepsilon$. This means $\|\sigma(0)\| < \varepsilon$ and $\|\sigma(2\pi)\| = \|T(\sigma(0))\| \le \|T\|\,\|\sigma(0)\| < 1\cdot\varepsilon = \varepsilon$. Therefore we can find a $\delta > 0$ such that $\phi \in [0,\delta[\ \cup\ ]2\pi - \delta, 2\pi]$ implies $\|\sigma(\phi)\| < \varepsilon$. Whence for $z \in \{e^{i\phi} : |\phi| < \delta\}$ we obtain $\|\hat{\sigma}(z)\| < \varepsilon$.

By (5.9) we obtain a bundle $p_T : \mathcal{E}_T \to S^1$ such that $\Gamma(p_T) = E_T$.

If we choose $F = \mathbb{R}$ with the usual norm and $T = -1$ (i.e. T is multiplication with -1), then $p_T : \mathcal{E}_T \to S^1$ is the Moebius strip.

If $F = \mathbb{R}^2$ with the Euclidean metric and if $T = \begin{pmatrix} 1 & 0 \\ 0 & -1 \end{pmatrix}$, then

$p_T : E_T \to S^1$ is homotopy equivalent to Klein's bottle.

More generally, if $F = \mathbb{R}^n$ with the Eukledian metric and if $T : \mathbb{R}^n \to \mathbb{R}^n$ is a linear operator with $\det(T) = -1$, then we obtain a higher dimensional analogon of the Moebius strip and Klein's bottle, resp.

It is well known, that for $F = \mathbb{R}^n$ and for invertible $T : \mathbb{R}^n \to \mathbb{R}^n$ the bundle $p_T : E_T \to S^1$ is locally trivial (see section 16 for definitions). Moreover, in this case we only have two isomorphism classes of locally trivial bundles over S^1, depending on whether $\det(T) > 0$, in which case we obtain the trivial bundle, or $\det(T) < 0$.

We shall see in section 17 that every bundle $p : E \to S^1$ whose stalks are of some fixed dimension $n < \infty$ and whose bundle space E is Hausdorff is in fact locally trivial and therefore isomorphic to one of the bundles $p_T : E_T \to S^1$, where $T : \mathbb{R}^n \to \mathbb{R}^n$ is invertible.

5.15 Sequence spaces. Let E be a topological vector space with a family of seminorms $(\nu_j)_{j \in J}$. Let $c_o(E)$ denote the vector space of all O-sequences in E equipped with the topology induced by the seminorms $(\vartheta_j)_{j \in J}$ given by

$$\vartheta_j((u_n)_{n \in \mathbb{N}}) = \sup \{\nu_j(u_n) : n \in \mathbb{N}\}.$$

It is well known that $c_o(E)$ is a closed subspace of $\prod_{n \in \mathbb{N}}^{\infty} E_n$, where $E_n = E$ for every $n \in \mathbb{N}$. Moreover, if we equip $\mathbb{N}$ with any topology finer than the cofinite topology (i.e. with any T_1-topology), then $c_o(E)$ is a function module, as multiplication with a bounded $\mathbb{K}$-valued function does not lead out of the class of O-sequences and as for every $\varepsilon > 0$, every $j \in J$ and every O-sequence $(u_n)_{n \in \mathbb{N}}$ in E the set $\{n \in \mathbb{N} : \nu_j(u_n) \geq \varepsilon\}$ is finite and thus closed.

Thus, we may construct a bundle $p : \mathcal{E} \to \mathbb{N}$ such that $c_o(E) \subset \Gamma(p)$, where $\mathcal{E} = \mathbb{N} \times E$ and where p is the first projection. The topology on $\mathcal{E}$ may be described as follows:

Let $(n_o, u_o) \in \mathcal{E} = \mathbb{N} \times E$. If we define a 0-sequence $(u_n)_{n \in \mathbb{N}} \in c_o(E)$ by $u_{n_o} = u_o$ and $u_n = 0$ for $u_n = 0$ for $n \neq n_o$, then this sequence gives a section of $p : \mathcal{E} \to \mathbb{N}$ passing through (n_o, u_o). Hence by (1.6(vii)) the sets of the form $\{(n,u) \in \mathcal{E} : n = n_o$ and $\nu_j(u - u_o) < \varepsilon$ or $n \neq n_o$, $n \in U$ and $\nu_j(u) < \varepsilon\}$, where U is an open neighborhood of n_o, $\varepsilon > 0$ and $j \in J$, form a neighborhood base at (n_o, u_o)

Finally, we calculate $\Gamma(p)$ in two special cases:

a) $\mathbb{N}$ carries the discrete topology. Then $\mathcal{E}$ carries the product topology of $\mathbb{N} \times E$ and therefore $\Gamma(p)$ consists of all bounded sequences with values in E.

b) $\mathbb{N}$ carries the cofinite topology. In this case we have $\Gamma(p) = c_o(E)$: Indeed, let $\sigma \in \Gamma(p)$. Then there is an $u_1 \in E$ such that $\sigma(1) = (1, u_1)$. As for a given $\varepsilon > 0$ and a given $j \in J$ the set

$$\mathcal{O} := \{(n,u) \in \mathcal{E} : n = 1 \text{ and } \nu_j(u - u_1) < \varepsilon \text{ or } n \neq 1 \text{ and } \nu_j(u) < \varepsilon\}$$

is an open neighborhood of $\sigma(1)$, we can find an open neighborhood U of 1 such that $\sigma(U) \subset \mathcal{O}$. As U is open in the cofinite topology, there is an $n_o \in \mathbb{N}$ such that $\{n : n \geq n_o\} \subset U$. This implies $\nu_j(\sigma(n)) < \varepsilon$ for all $n \geq n_o$, i.e. $\sigma \in c_o(E)$.

5.16 Example. Let $X = [0,1]$ be the unit interval with its usual topology and let E be the completion of $C([0,1])$ under the norm $|||\cdot|||$ given by

$$|||f||| = \max \{|f(0)|, \sup_{0<r\leq 1} r\cdot|f(r)|\} .$$

It is easy to see that E is a C([0,1])-submodule of $\prod_{0\leq r\leq 1}^{\infty} E_r$, where E_r is identical with $\mathbb{R}$ equipped with the norm

$$|||\alpha|||_r := \begin{cases} |\alpha| & r = 0 \\ r\cdot|\alpha| & r \neq 0 \end{cases}$$

Furthermore, E is indeed a function module with stalks E_r and base space X. Hence , by (5.9) there is a bundle $p : E \to 0,1$ such that E is isometrically isomorphic with $\Gamma(p)$. We may identify the <u>set</u> E with $[0,1]\times\mathbb{R}$. If we do so, the mapping p becomes the first projection.

Let us try to describe the topology of E. The open sets are given by tubes and by (1.6(vii)) we may use tubes around constant sections. Thus, let $(r,\alpha) \in [0,1]\times\mathbb{R}$ be an element of the bundle space. If $r \neq 0$, then a neighborhood base at (r,α) is given by sets of the form

$$\{(s,\beta) : |r - s| < \varepsilon, s\cdot|\alpha - \beta| < \varepsilon\}, \quad 0 < \varepsilon < r/2$$

and it turns out that the subset $]0,1]\times\mathbb{R} \subset E$ carries the usual product topology.
A neighborhood base at $(0,\alpha) \in [0,1]\times\mathbb{R}$ is given by

$$\{(0,\beta) : |\alpha - \beta| < \varepsilon\} \cup \{(s,\beta) : |s| < \varepsilon, s\cdot|\alpha - \beta| < \varepsilon\}$$

Hence every open neighborhood of $(0,\alpha)$ contains the elements of the form $(s,0)$, if we only choose $s < \frac{\varepsilon}{2\cdot|\alpha|}$, and we conclude that the closure of the set $[0,1]\times\{0\} \subset E$ is equal to $\{0\}\times\mathbb{R} \cup [0,1]\times\{0\}$ and that the closure of $\{\alpha \in E : |||\alpha||| \leq 1\}$ is equal to the set $\{0\}\times\mathbb{R} \cup \{\alpha \in E : |||\alpha||| \leq 1\}$. This provides us with an example in which the "unit ball" $\{\alpha \in E : |||\alpha||| \leq 1\}$ is not closed.

We leave it as an exercise for the reader to verify that

$$\Gamma(p) = \{f : [0,1] \to \mathbb{R} : \lim_{r \to s} r \cdot f(r) = s \cdot f(s)\}$$

i.e. f belongs to $\Gamma(p)$ if and only if the mapping $f : [0,1] \to \mathbb{R}$ is continuous at every $r \neq 0$ and satisfies the equation $\lim_{r \to 0} r \cdot f(r) = 0$. Especially, the mapping $\chi_{\{0\}}$ defined by

$$\chi_{\{0\}} = \begin{cases} 0 & \text{if } s \neq 0 \\ 1 & \text{if } s = 0 \end{cases}$$

is an element $\Gamma(p)$.

We conclude this section with the remark that also all weighted vector valued function spaces in the sense of Bierstedt, Kleinstück, Machado, Nachbin, Prolla et al. fall under the notion of function modules. For a precise definition and a treatment of these examples we refer to the papers of the authors just mentioned.

6. Some algebraic aspects of Ω-spaces

In this section we collect some properties of Ω-spaces which will be needed later on. Nothing new will be found here, all the results are folklore. Therefore we can confidently leave all the proofs to the reader.

6.1 Definition. Let E and F be two Ω-spaces of type $\tau : I \to \mathbb{N}$.
(i) An *Ω-morphism* from E into F is a linear map $\phi : E \to F$ such that for all $i \in I$ and all $a_1, \ldots, a_{\tau(i)} \in E$ we have

$$f_i(\phi(a_1), \ldots, \phi(a_{\tau(i)})) = \phi(f_i(a_1, \ldots, a_{\tau(i)})).$$

(ii) A linear subspace $N \subset E$ is called an *Ω-ideal*, if for all $i \in I$ and all sequences of ordered pairs $(a_1, b_1), \ldots, (a_{\tau(i)}, b_{\tau(i)}) \in E \times E$ the relations $a_j - b_j \in N$ for all $1 \le j \le \tau(i)$ imply $f_i(a_1, \ldots, a_{\tau(i)}) - f_i(b_1, \ldots, b_{\tau(i)}) \in N$. □

If E is a vector lattice, then the Ω-ideals are exactly the ideals in the usual sense; if E is an algebra over $\mathbb{K}$, then the Ω-ideals are also the usual algebra ideals.

6.2 Proposition. *Let $\phi : E \to F$ be an Ω-morphism between two Ω-spaces E and F. Then* $\ker \phi := \phi^{-1}(0)$ *is an Ω-ideal.* □

The next result states a kind a reverse of (6.2):

6.3 Proposition. *Let E be a (topological) Ω-space of type τ and let N be a (closed) Ω-ideal of E. Then E/N is also a (topological) Ω-space of type τ, where the additional operations $(f_i)_{i \in I}$ on E/N* *are defined by*

$$f_i(a_1 + N,\dots,a_{\tau(i)} + N) := f_i(a_1,\dots,a_{\tau(i)}) + N.$$

Moreover, the canonical quotient map $\pi : E \to E/N$ *is a (continuous and open)* Ω*-morphism.*

Conversely, if $\psi : E \to F$ *is a (continuous and open) quotient map and an* Ω*-morphism and if* $N = \ker\psi$*, then the* Ω*-spaces* E/N *and* F *are (topologically) isomorphic.* □

6.4 Proposition. *Let* $\psi : E \to F$ *be a (continuous)* Ω*-morphism between the* Ω*-spaces* E *and* F*, let* $N = \ker\psi$ *and let* $\pi : E \to E/N$ *be the canonical quotient map. Then there exists an injective (and continuous)* Ω*-morphism* $\phi : E/N \to F$ *such that* $\psi = \phi\circ\pi$. □

6.5 Proposition. *Let* E *be a topological* Ω*-space. Then the closure of an* Ω*-ideal is again an* Ω*-ideal.* □

6.6 Proposition. *Let* E *be a (topological)* Ω*-space and let* $(N_\lambda)_{\lambda\in\Lambda}$ *be a family of (closed)* Ω*-ideals. Then* $\bigcap_{\lambda\in\Lambda} N_\lambda$ *and* $\sum_{\lambda\in\Lambda} N_\lambda$ *(resp.* $\overline{\sum_{\lambda\in\Lambda} N_\lambda}$*) are again (closed)* Ω*-ideals. Especially, the (closed)* Ω*-ideals form a complete sublattice of the complete lattice of all (closed) subspaces.* □

6.7 Notation. With $Id_\Omega(E)$ we denote the complete lattice of all <u>closed</u> Ω-ideals of a topological Ω-space E.

7. A third description of spaces of sections : C(X)-convex modules

We know from (1.6) that for every bundle $p : E \to X$ the space of all sections is a $C(X)$-module. Now suppose that somebody gave us a $C(X)$-module E. Then we ask ourselves: Is it possible to "spread E continuously across X", i.e. is there a bundle $p : E \to X$ such that E and $\Gamma(p)$ are isomorphic? In this section we shall describe those $C(X)$-modules for which this is possible.

If we are dealing with Ω-spaces, we also have to worry about the additional operations. Hence we note the following: If $p : E \to X$ is a bundle of Ω-spaces over a quasicompact base space, then $\Gamma(p)$ is a topological Ω-space. Moreover, if $A \subset X$ is a subset, then the set $N_A := \{\sigma \in \Gamma(p) : \sigma_{/A} = 0\}$ is a closed Ω-ideal. Especially, if $f \in C(X)$ is a continuous $\mathbb{K}$-valued function on X and if we let $A = f^{-1}(\mathbb{K} \setminus \{0\})$, then $f^{\perp} := N_A = \{\sigma \in \Gamma(p) : f\cdot\sigma = 0\}$ is an Ω-ideal. This leads to the following definition:

7.1 Definition. A $C_b(X)$-Ω-*module* E is a topological Ω-space which is at the same time a $C_b(X)$-module such that

(i) the multiplication $(f,a) \to f\cdot a : C_b(X) \times E \to E$ is continuous.

(ii) for every element $f \in C_b(X)$ the set $f^{\perp} := \{a \in E : f\cdot a = 0\}$ is an Ω-ideal. □

Note that by the continuity of the multiplication the Ω-ideal $f^{\perp}$ is automatically closed for every $f \in C_b(X)$.

For the following results see also the work of J.Varela ([Va 75]) in the case of Banach spaces and the paper [Mö 78] of H.Möller.

As it is always convenient to work with compact Hausdorff spaces instead of arbitrary topological spaces, we shall heavily make use of the following fact:
For every topological space X there is a compact (Hausdorff) space βX (the Stone-Čech-compactification of X) and a continuous mapping $i : X \to \beta X$ such that

(i) $i(X)$ is dense in βX.

(ii) the mapping $f \to f \circ i : C(\beta X) \to C_b(X)$ is a bijection preserving the sup-norm and the algebraic structure.

This means that in the proofs of many results concerning C(X)-modules we may assume w.l.o.g. that the space X is compact. This will be at least be possible as long as we do not talk about points of X.

7.2 Proposition. *Let* E *be a* $C_b(X)$-Ω-*module and let* I *be an ideal of* $C_b(X)$. *Then the closure of the complex product* $I \cdot E$ *is an* Ω-*ideal. Moreover, if* X *is compact and if* I *is a closed ideal, then* $I = \{f : f_{/A} = 0\}$ *for some closed subset* $A \subset X$. *In this case, if we let* $F_A := \{f \in C_b(X) : f^{-1}(0)$ is a neighborhood of $A\}$, *then we have the following equalities:*

$$\overline{I \cdot E} = \overline{F_A \cdot E}$$

$$= \overline{\Sigma \{f^{\perp} : f_{/A} = 1\}}$$

$$= \overline{\cup \{f^{\perp} : f_{/U} = 1 \text{ for some open set } U \supset A\}}$$

$$= \overline{\cup \{f^{\perp} : f_{/A} = 1\}}.$$

Moreover, in all these cases we may restrict ourselves to continuous functions $f : X \to [0,1] \subset \mathbb{K}$.

Proof. By the continuity of the multiplication with elements in C(X) we may assume that I is closed. Moreover, from (7.1(ii)) we know that $f^{\perp}$ is an Ω-ideal and therefore by (6.6) the closure of $\Sigma\{f^{\perp} : f_{/A} = 1\}$ is an Ω-ideal. Hence by the above remarks concerning the Stone-Čech compactification it is enough to prove the assertions for compact X in (7.2).

Hence we assume that X is compact and that $A \subset X$ is a closed subset. As F_A is dense in $I = \{f : f_{/A} = 0\}$, we obtain $\overline{I\cdot E} = \overline{F_A\cdot E}$.

Let $f \in F_A$. then there is an open neighborhood U of A such that $f_{/U} = 0$. Let V be an open set such that $A \subset V \subset \overline{V} \subset U$ and choose a continuous function $g : X \to [0,1]$ such that $g(V) = \{1\}$ and $g(X \setminus U) = \{0\}$. Then $g\cdot f = 0$ and whence for all $a \in E$ we have $g\cdot(f\cdot a) = 0$, i.e. $f\cdot E \subset g^{\perp}$. This yields the inclusion

$$\overline{F_A\cdot E} \subset (\cup\{f^{\perp} : f_{/U} = 1 \text{ for some open set } U \supset A\})^{-}$$

Obviously, we have

$$(\cup\{f^{\perp} : f_{/U} = 1 \text{ for some open set } U \supset A\})^{-} \subset (\Sigma\{f^{\perp} : f_{/A} = 1\})^{-}$$

Further, let $a \in \Sigma\{f^{\perp} : f_{/A} = 1\}$. Then there are elements $f_1,\dots,f_n \in C(X)$ and $a_1,\dots,a_n \in E$ such that $f_{i/A} = 1$, $f_i\cdot a_i = 0$ and $a_1 + \dots + a_n = a$. Let $g = f_1\cdot\ldots\cdot f_n$. Then $g_{/A} = 1$ and $g\cdot a_i = 0$ for all $1 \le i \le n$. Thus, we obtain $g\cdot a = 0$, i.e. $a \in g^{\perp}$. Hence we obtain the inclusion

$$(\Sigma\{f^{\perp} : f_{/A} = 1\})^{-} \subset (\cup\{f^{\perp} : f_{/A} = 1\})^{-}.$$

Finally, let $a \in \cup\{f^{\perp} : f_{/A} = 1\}$. Then $f\cdot a = 0$ for some $f \in C(X)$ with $f_{/A} = 1$. Let $g = 1 - f$. Then g vanishes on A and therefore belongs to the ideal I. Moreover, we have $g\cdot a = (1 - f)\cdot a = a - f\cdot a = a$, i.e. $a \in I\cdot E$. As this implies

$$(\cup\{f\ \ : f_{/A} = 1\})^{-} \subset \overline{I\cdot E}$$

our proof is complete. □

7.3 Corollary. *Let* E *be a* $C_b(X)$-Ω*-module and let* $A \subset X$ *be a subset. If we let* $I = \{f : f_{/A} = 0\}$, *then*

$$\overline{I\cdot E} \subset \overline{\cup \{f^{\perp} : f_{/U} = 1 \text{ for some open set } U \supset A \text{ and } 0 \le f \le 1\}}$$

Proof. This assertion follows for compact X immediatly from (7.2). If X is not compact, consider that Stone-Čech compactification βX of X and let $i : X \to \beta X$ be the canonical map. Then for $f \in C_b(X)$ we have $f(A) = \{0\}$ if and only if there is a $f' \in C(\beta X)$ with $f = f' \circ i$ and $f'(i(A)) = 0$. Moreover, if $f'_{/U} = 1$ for some $f' \in C(\beta X)$ and some open set $U \subset \beta X$, then for $f = f' \circ i$ we have $f_{/i^{-1}(U)} = 1$ and $i^{-1}(U)$ is also open. Now some straightforward arguments complete the proof. □

7.4 Proposition. *Let* E *be a* $C_b(X)$-Ω*-module. Then we have:*

(i) $0 \in E$ *has a neighborhood base consisting of closed, convex and circled sets* $A \subset E$ *such that* $f\cdot A \subset A$ *for all* $f \in C_b(X)$ *with* $\|f\| \le 1$.

(ii) *The topology on* E *is generated by a family* $(\nu_j)_{j\in J}$ *of seminorms satisfying* $\nu_j(f\cdot a) \le \|f\|\cdot\nu_j(a)$ *for all* $j \in J$.

Proof. Let ν be the gauge function of a closed, convex and circled neighborhood A of 0 fullfiling $f\cdot A \subset A$ for all $f \in C_b(X)$ with $\|f\| \le 1$. Then it is easy to check that $\nu(f\cdot a) \le \|f\|\cdot\nu(a)$. Whence (i) implies (ii). It remains to check (i):
Let U be any closed convex and circled neighborhood of 0. Then, by the continuity of multiplication by $f \in C_b(X)$, we can find an $\varepsilon > 0$ and a closed, convex and circled neighborhood V' of 0 such that $\|f\| \le \varepsilon$ and $v \in V'$ imply $f\cdot v \in U$. Let $V := \varepsilon\cdot V'$. Then V is still a

closed, convex, circled neighborhood of 0 and we have $f \cdot V \subset U$ for all $f \in C_b(X)$ with $\|f\| \leq 1$. Let

$$W := \cup \{f \cdot V : \|f\| \leq 1\}.$$

From $1 \cdot V = V$ we obtain $V \subset W \subset U$. Let A be the closed, convex, circled hull of W. Then A is a closed, convex and circled neighborhood of 0 contained in U. Moreover, $f \cdot W \subset W$ for all $f \in C_b(X)$ with $\|f\| \leq 1$ and the continuity of multiplication imply $f \cdot A \subset A$ for all $f \in C_b(X)$ with $\|f\| \leq 1$. This completes the proof. □

The next lemma, due to J.Varela [Va 75] in the case of Banach spaces and due to H.Möller in our setting, opens the door to a connection between $C(X)$-modules and bundles:

7.5 Lemma (Varela). *Let* E *be a* $C_b(X)$*-module and let* ν *be a continuous seminorm on* E *satisfying* $\nu(f \cdot a) \leq \|f\| \cdot \nu(a)$ *for all* $a \in E$ *and all* $f \in C_b(X)$. *If* $A \subset X$ *is any subset and if* $I = \{f \in C_b(X) : f_{/A} = 0\}$, *then for all* $a \in E$ *we have*

$$\begin{aligned} \nu(a + \overline{I \cdot E}) &:= \inf \{\nu(a + b) : b \in \overline{I \cdot E}\} \\ &= \inf \{\nu(f \cdot a) : f \in C_b(X),\ 0 \leq f \leq 1 \text{ and } A \subset f^{-1}(\{1\})^{o}\}. \end{aligned}$$

Proof. Let

$$\begin{aligned} l &:= \inf \{\nu(a + b) : b \in \overline{I \cdot E}\} \\ r &:= \inf \{\nu(f \cdot a) : f \in C_b(X),\ 0 \leq f \leq 1 \text{ and } A \subset f^{-1}(\{1\})^{o}\}. \end{aligned}$$

If $f : X \to \mathbb{K}$ is a continuous and bounded function with constant value 1 on A, then $1 - f$ vanishes on A and thus belongs to I. Thus, we know that $(1 - f) \cdot a$ belongs to $I \cdot E$ for every $a \in E$. As we may write $f \cdot a = a + (f - 1) \cdot a$, we obtain $l \leq r$.

Conversely, let $\varepsilon > 0$. By (7.3) and the definition of l we can find an $f \in C_b(X)$ with $0 \leq f \leq 1$ and $A \subset f^{-1}(\{1\})^{\circ}$ and an element $b \in f^{\perp}$ such that $\nu(a + b) < 1 + \varepsilon$. This implies $r \leq \nu(f \cdot a) = \nu(f \cdot a + f \cdot b) = \nu(f \cdot (a + b)) \leq \|f\| \cdot \nu(a + b) \leq 1 + \varepsilon$. As $\varepsilon > 0$ was arbitrary, this yields $r \leq 1$. □

Before we proceed, let us introduce some notations:
Let E again be a $C_b(X)$-Ω-module, let ν be a continuous seminorm on E, let $a \in E$ and let $x \in X$. We define

$$I_x := \{f \in C_b(X) : f(x) = 0\}$$

$$N_x := \overline{I_x \cdot E}$$

$$E_x := E/N_x$$

$\varepsilon_x : E \to E_x$ is the canonical projection

$$a_x := \varepsilon_x(a)$$

$$\nu^x(a_x) := \inf \{\nu(a + b) : b \in N_x\}$$

$$N := \bigcap_{x \in X} N_x$$

By (7.2) the subspace N_x is an Ω-ideal. Hence, by (6.3), the quotient space E_x is a topological Ω-space, too, and the quotient map ε_x is an Ω-morphism.
Moreover, we have

7.6 $\quad \varepsilon_x(f \cdot a) = f(x) \cdot a_x$ for all $a \in E$ and all $f \in C_b(X)$.
(Indeed, $f(x) = 0$ implies $f \cdot a \in N_x$, whence $\varepsilon_x(f \cdot a) = 0$. Now let $f \in C_b(X)$ be arbitrary. Define a continuous and bounded function $g \in C_b(X)$ by $g = f - f(x) \cdot 1$. Then g vanishes at the point x and we may compute: $\varepsilon_x(f \cdot a) = \varepsilon_x((f - f(x) \cdot 1) \cdot a + f(x) \cdot a) = \varepsilon_x(g \cdot a + f(x) \cdot a) = \varepsilon_x(g \cdot a) + f(x) \cdot \varepsilon_x(a) = f(x) \cdot a_x$.)

7.7 If we fix an element $a \in E$ and if $\nu(f \cdot b) \leq \|f\| \cdot \nu(b)$ for all $f \in C_b(X)$ and all $b \in E$, then the mapping

$$\nu^-(a_-) : X \to \mathbb{R}$$
$$x \to \nu^x(a_x)$$

is upper semicontinuous.
(Let $\nu^x(a_x) < M$. By (7.5) there is an open neighborhood U of x and a continuous mapping $f : X \to [0,1] \subset \mathbb{R}$ such that $f_{/U} = 1$ and such that $\nu(f\cdot a) < M$. Using (7.5) again, we conclude that $\nu^y(a_y) < M$ for all $y \in U$.)

It is now obvious from (7.4) and (7.7) that the Ω-space E/N may algebraically be embedded into $\prod^{\infty}_{x\in X} E_x$ and that this embedding is continuous. Moreover, under this embedding, the closure of E/N in $\prod^{\infty}_{x\in X} E_x$ is an Ω-function module with base space X and stalks E_x. Further, the stronger axiom (FFM3) is automatically satisfied. Thus, we may state:

7.8 Proposition. *Let* E *be a* $C_b(X)$*-module. Then there is a full bundle* $p : E \to X$ *and a continuous and injective* Ω*-morphism* $i : E/N \to \Gamma(p)$. *In addition, this bundle has the property that for every* $\alpha \in E$ *there is an element* $u \in E/N$ *with* $i(u)(p(\alpha)) = \alpha$. *If* E *is a* $C_b(X)$*-*Ω*-module and if* X *is completely regular, then this bundle* $p : E \to X$ *is in fact a bundle of* Ω*-spaces.* □

In order to present the whole space E as a space of sections, we need two things : Firstly, the mapping $i : E/N \to \Gamma(p)$ given in (7.8) should be open onto its image. Secondly, the subspace $N \subset E$ should be trivial. Unfortunatly, the following example shows that it may happen that $N = E$, even if E is a Banach space and even if the base space X is compact:

7.9 Example. Let $X = [0,1]$ be the unit interval with its usual topology and let E be the completion of C([0,1]) in the norm given by

$$|||f||| = \int_0^1 |f(x)|dx .$$

Then E is a C([0,1])-module and we have $|||f \cdot m||| \le ||f|| \cdot |||m|||$ for all $f \in C([0,1])$ and all $m \in E$. In this case, $N_x = E$ for all $x \in [0,1]$. Indeed, let $x_0 \in [0,1]$, let $m \in C([0,1]) \subset E$ and let $\varepsilon > 0$. If we define $\delta = \frac{\varepsilon}{2 \cdot |||m|||}$, we may find an element $f \in C([0,1])$ such that $0 \le f \le 1$, $f(x_0) = 0$ and $f(x) = 1$ for all x with $|x - x_0| > \delta$. Then the element $f \cdot m$ will belong to N_x and we have

$$\begin{aligned} |||m - f \cdot m||| &= |||(1 - f) \cdot m||| \\ &= \int_0^1 (1 - f(x)) \cdot |m(x)| \, dx \\ &\le 2 \cdot \delta \cdot |||m||| \\ &< \varepsilon \end{aligned}$$

This yields $m \in N_x$. As C([0,1]) is dense in E, we conclude that $E = N_x$.

Suppose that our C(X)-module E may be represented as a C(X)-module of sections in a bundle $p : \mathcal{E} \to X$. From (1.6.(x)) we then may deduce that E is locally C(X)-convex in the sense of the following definition:

7.10 Definition. Let E be a $C_b(X)$-module.

(i) A subset $A \subset E$ is called C(X)-*convex*, if for all $m,n \in A$ and all $f \in C_b(X)$ with $0 \le f \le 1$ we have $f \cdot m + (1 - f) \cdot n \in A$.

(ii) The $C_b(X)$-module E is said to be *locally* C(X)-*convex*, provided that $0 \in E$ has a neighborhood base of C(X)-convex sets. □

Note that for every $C(X)$-convex subset $A \subset E$ the convex and circled closed hull of A is also $C(X)$-convex. Hence E is locally $C(X)$-convex if and only if 0 has a neighborhood base of closed, convex and circled $C(X)$-convex subsets. If we pass to the gauge functions given by these sets, we obtain:

7.11 Proposition. *A* $C_b(X)$*-module* E *is* $C(X)$*-convex if and only if the topology on* E *is induced by a family of seminorms* $(\nu_j)_{j \in J}$ *satisfying the following condition*

If $m,n \in E$ with $\nu_j(m),\ \nu_j(n) \leq 1$ and if $f \in C_b(X)$ with $0 \leq f \leq 1$, then $\nu_j(f \cdot m + (1 - f) \cdot n) \leq 1$. □

Hence we are once again led to a closer look on seminorms in $C(X)$-modules. Let us start with two lemmata:

7.12 Lemma. *Let* E *be* $C_b(X)$*-module and let* ν *be a seminorm on* E *satisfying* $\nu(f \cdot m) \leq \|f\| \cdot \nu(m)$ *for all* $m \in E$ *and all* $f \in C_b(X)$. *Then* $|f| \leq |g|$ *implies* $\nu(f \cdot m) \leq \nu(g \cdot m)$ *for all* $m \in E$, $f,g \in C_b(X)$.

Proof. Suppose that $g(x) \neq 0$ for all $x \in X$. Then we may compute: $\nu(f \cdot m) = \nu(g \cdot \frac{f}{g} \cdot m) \leq \|\frac{f}{g}\| \cdot \nu(g \cdot m) \leq \nu(g \cdot m)$.
The proof of the general case is a modification of this idea: Firstly, we may assume without loss of generality that X is compact. Let $\varepsilon > 0$ and let $A := \{x : |g(x)| \geq \varepsilon\}$. Then A is closed in X. For every $x \in A$ we define $h(x) := \frac{f(x)}{g(x)}$. Clearly, the mapping $h : A \to \mathbb{K}$ is continuous and we have $|h(x)| \leq 1$ for all $x \in A$. Let $\overline{h} : X \to \mathbb{K}$ be an extension of h with $|\overline{h}| \leq 1$. Then an easy calculation shows that $\|h \cdot g - f\| < 2 \cdot \varepsilon$. This implies $\nu(f \cdot m) \leq \nu((f - g \cdot h) \cdot m) + \nu(h \cdot g \cdot m) \leq 2 \cdot \varepsilon \cdot \nu(m) + \nu(g \cdot m)$. As $\varepsilon > 0$ was arbitrary, we obtain $\nu(f \cdot m) \leq \nu(g \cdot m)$. □

The next lemma is due to Bohnenblust and Kakutani (see [BK 41]):

7.13 Lemma. *Let* ν *be a seminorm on* $C_b(X)$ *satisfying* $\nu(f\cdot g) \leq \|f\| \cdot\nu(g)$. *If* $\nu(f\vee g) = \max\{\nu(f),\nu(g)\}$ *whenever* $f\wedge g = 0$, *then we have* $\nu(f\vee g) = \max\{\nu(f),\nu(g)\}$ *for all* $0 \leq f,g \in C_b(X)$.

Proof. Again, we may assume without loss of generality that X is compact. Let $I = \{f \in C(X) : \nu(f) = 0\}$. Then I is an ideal of C(X) and ν induces a norm on C(X)/I. Moreover, C(X)/I is a vector lattice and we have $|f + I| = |f| + I$ for all $f \in C(X)$. From (7.12) we conclude that $\nu(f) = \nu(|f|)$. This implies the equation $\nu(|f + I|) = = \nu(|f|) = \nu(f) = \nu(f + I)$. Hence the space C(X)/I is a normed vector lattice.

Let $(f + I) \wedge (g + I) = 0 = (f\wedge g) + I$. Then $f\wedge g$ belongs to the ideal I. Substituting f and g by $f - f\wedge g$ and $g - f\wedge g$ resp., we may assume that $f\wedge g = 0$. Using our hypothesis, we may compute: $\nu((f + I)\vee(g + I)) = \nu(f\vee g) = \max\{\nu(f + I),\nu(g + I)\}$. Let E be the completion of C(X)/I in the norm ν. Some standard arguments show that E still satisfies $\nu(a\vee b) = \max\{\nu(a),\nu(b)\}$ whenever $a\wedge b = 0$. Now we deduce from [BK 41] the E is an abstract M-space, i.e. E satisfies the equation $\nu(a\vee b) = \max\{\nu(a),\nu(b)\}$ for all $a,b \geq 0$.

After these preparations it is easy to show (7.13): For all $f,g \in C_b(X)$ with $f,g \geq 0$ we have $\nu(f\vee g) = \nu(f\vee g + I) = = \nu((f + I)\vee(g + I)) = \max\{\nu(f + I),\nu(g + I)\} = \max\{\nu(f),\nu(g)\}$. □

The following result is due to several authors: The equivalence of (1), (2) and (5) may be found in [Ho 75]. R.A. Bowshell [Bo 75] showed that (3) and (5) are equivalent and raised the question whether (3) and (4) are the same. In the present form, the next proposition is once again due to H.Möller:

7.14 Proposition. *Let E be a $C_b(X)$-module and let ν be a seminorm on E satisfying $\nu(f\cdot m) \le \|f\|\cdot\nu(m)$. Then the following conditions are equivalent:*

(1) If $f \in C_b(X)$ with $0 \le f \le 1$ and if $m,n \in E$ with $\nu(m),\nu(n) \le 1$, then we have also $\nu(f\cdot m + (1-f)\cdot n) \le 1$.

(2) If $0 \le f,g \in C_b(X)$ and if $m,n \in E$, then we have $\nu(f\cdot m + g\cdot n) \le \|f+g\|\cdot \max\{\nu(m),\nu(n)\}$.

(3) If $f,g \in C_b(X)$ with $f\cdot g = 0$ and if $m \in E$, then $\nu((f+g)\cdot m) = \max\{\nu(f\cdot m),\nu(g\cdot m)\}$.

(4) If $0 \le f,g \in C_b(X)$ and if $m \in E$, then we have $\nu((f\vee g)\cdot m) = \max\{\nu(f\cdot m),\nu(g\cdot m)\}$.

(5) For every $m \in E$ the following equation is true:
$\nu(m) = \sup\{\nu(m + \overline{I\cdot E}) : I$ is a maximal closed ideal of $C_b(X)\}$.

If X is quasicompact, then these conditions are also equivalent to

(5') If $m \in E$, then $\nu(m) = \sup\{\nu^x(m_x) : x \in X\}$.

Proof. The implication (2) → (1) is trivial.

(1) → (3): Assume that $f\cdot g = 0$. Then we have $|f+g| = |f| + |g| \ge |f|$. Thus (7.12) gives us the inequality $\nu(f\cdot m) \le \nu((f+g)\cdot m)$, i.e. $\max\{\nu(f\cdot m),\nu(g\cdot m)\} \le \nu((f+g)\cdot m)$.

Conversely, suppose that $\nu(f\cdot m),\nu(g\cdot m) \le 1$. We have to show that $\nu((f+g)\cdot m) \le 1 + \nu(m)\cdot\varepsilon$, where $\varepsilon > 0$ is arbitrary.

Thus, let $\varepsilon > 0$ and let $A = \{x \in X : f(x) = 0\}$ and $B = \{x \in X : |f(x)| \ge \varepsilon\}$. Choose any continuous mapping $h : X \to [0,1]$ with $h(A) = \{0\}$ and $h(B) = \{1\}$. (Here we again made the assumption that X is compact, which is possible w.l.o.g.) Then we have $\|f - h f\| \le \varepsilon$ and $(1-h)\cdot g = g$, because $g(x) \ne 0$ implies $x \in A$ and hence $(1-h)(x) = 1$. We now conclude

$$\nu((f + g)\cdot m) = \nu(h\cdot f\cdot m + (1 - h)\cdot g\cdot m + (f - h\cdot f)\cdot m)$$
$$\leq \nu(h\cdot f\cdot m + (1 - h)\cdot g\cdot m) + \|f - h\cdot f\|\cdot\nu(m)$$
$$\leq 1 + \nu(m)\cdot\varepsilon.$$

(3) → (4): Let $m \in E$. Define a seminorm ν_m on $C_b(X)$ by $\nu_m(f) := \nu(f\cdot m)$. Now apply (7.13)

(4) → (5),(5'): The maximal closed ideals of $C_b(X)$ correspond to the maximal closed ideals of $C(\beta X)$. Hence we may assume w.l.o.g. that X is quasicompact. In this case the maximal ideals of $C_b(X)$ are of the form $I_x = \{f \in C(X) : f(x) = 0\}$.Thus, it is enough to prove (5').
Obviously, we have $\sup \{\nu^x(m_x) : x \in X\} \leq \nu(m)$.
Conversely, let $\varepsilon > 0$ and assume that $\sup \{\nu^x(m_x) : x \in X\} < \nu(m) - \varepsilon$. Applying (7.5), we find for every point $x \in X$ an open neighborhood V_x and a continuous function $f_x : X \to [0,1]$ with $f_x(V_x) = \{1\}$ such that $\nu(f_x\cdot m) < \nu(m) - \varepsilon$. As X is quasicompact, there is a finite number of point $x_1,\ldots,x_n \in X$ such that $V_{x_1} \cup \ldots \cup V_{x_n} = X$. As this implies $f_{x_1} \vee \ldots \vee f_{x_n} = 1$, we obtain from (4) the inequality

$$\nu(m) = \nu((f_{x_1} \vee \ldots \vee f_{x_n})\cdot m)$$
$$= \max \{\nu(f_{x_i}\cdot m) : 1 \leq i \leq n\}$$
$$< \nu(m) - \varepsilon$$

a contradiction.

(5),(5') → (2): Again, we may assume that X is compact and hence it is enough to show that (5') implies (2). Firstly, note that $(f\cdot m)_x = f(x)\cdot m_x$ by (7.6). This yields $\nu^x((f\cdot m)_x + (g\cdot n)_x) \leq$ $\leq |f(x) + g(x)|\cdot\max \{\nu^x(m_x),\nu^x(n_x)\}$ whenever $f(x), g(x) \geq 0$. Now an easy calculation using (5') shows (2). □

If we combine (7.11) and (7.14), we obtain

7.15 Proposition. *Let* X *be a quasicompact space and let* E *be a* C(X)*-module. If* E *is locally* C(X)*-convex, then* $N = \cap \{\overline{I_x \cdot E} : x \in X\} = 0$.

Proof. The topology on E is induced by a family of seminorm $(\nu_j)_{j\in J}$ which satisfy condition (1) of (7.14). Let $m \in N$. Then $m_x = 0$ for all $x \in X$. Hence for all $j \in J$ and all $x \in X$ we have $\nu_j^x(m_x) = 0$. From (7.14) we deduce that $\nu_j(m) = 0$ for all $j \in J$, i.e. $m = 0$. □

We now come to a central result, which appears in different form already in the work of Nachbin (see [Na 59]). In the present form however, this theorem is due to K.H.Hofmann for Banach spaces and to H.Möller in the general case:

7.16 Theorem. *Let* X *be a quasicompact space and let* E *be a* C(X)*-module. Then* E *is locally* C(X)*-convex if and only if* E *is (topologically and algebraically isomorphic to) a* C(X)*-submodule of* $\Gamma(p)$, *where* $p : \mathcal{E} \to X$ *is a bundle.*
Moreover, if $\alpha \in \mathcal{E}$, *then there is an element* σ *in (the image of)* E *such that* $\sigma(p(\alpha)) = \alpha$.
If X *is compact, then* E *is dense in* $\Gamma(p)$. *Hence, if* E *is complete, then* $E \simeq \Gamma(p)$.

Proof. Every C(X)-submodule E of $\Gamma(p)$ is locally C(X)-convex by (1.6.(x)). The other direction follows from (7.8) and (7.15); the last statement is a consequence of the Stone-Weierstraß theorem (4.2). □

For C(X)-Ω-modules, we deduce from (7.8) and (7.16) the following

7.17 Complement. *Let* E *be an* Ω*-space which is at the same time a* C(X)*-module for a certain compact space* X. *Then* E *is a locally* C(X)*-convex* C(X)-Ω*-module if and only if* E *is (topologically and algebraically isomorphic to) a dense* C(X)*-submodule of the* Ω*-space* $\Gamma(p)$, *where* $p : E \to X$ *is a bundle of* Ω*-spaces.* □

Let us state some corollaries which will cover the most important cases:

7.18 Definition. A normed C(X)-module E is called *locally* C(X)*-convex*, if for all $f \in C_b(X)$ with $0 \le f \le 1$ and all $m,n \in E$ with $||m||$, $||n|| \le 1$ we have $||f\cdot m + (1 - f)\cdot n|| \le 1$. □

7.19 Corollary. *Let* X *be a compact space and let* E *be a Banach space which is a* C(X)*-module. Then* E *is locally* C(X)*-convex if and only if there is a bundle* $p : E \to X$ *of Banach spaces such that* E *is isometrically isomorphic to* $\Gamma(p)$. □

In section 14, notably (14.11), we shall see that (up to isomorphy) the bundle $p : E \to X$ given in (7.19) is also unique. On the other hand, the space X may be to "large" for E in the sense that "many" stalks of the bundle $p : E \to X$ are 0. This happens for instance, if we define a multiplication on E with elements of C(X) by $f\cdot m = f(x_o)\cdot m$ for a fixed $x_o \in X$. It is then easy to see that all the stalks of the bundle $p : E \to X$ will be equal to 0, except for the stalk over x_o, which will be equal to E itself. Hence we may as well choose the smaller space $\{x_o\}$ for a base space of the bundle $p : E \to X$ without losing any information. This leads us to the following observations:

7.20 If E is any $C_b(X)$-module, we define

$$E^{\perp} := \{f \in C_b(X) : f \cdot a = 0 \text{ for all } a \in E\}$$

It is clear that $E^{\perp}$ is a closed ideal of $C(X)$. If X is compact, then there is a closed subset $A \subset X$ such that $E^{\perp} = \{f \in C(X) : f_{/A} = 0\}$ and $C(X)/E^{\perp} \simeq C(A)$. Obviously, E is also a $C(A)$-module. Hence we may in all cases replace the compact space X by the smaller set A.

7.21 Definition. (i) A $C_b(X)$-module E is called *reduced* if $f \cdot a = 0$ for all $a \in E$ implies $f = 0$.

(ii) A bundle $p : E \to X$ is called *reduced* if $\{x \in X : p^{-1}(x) \neq 0\}$ is dense in X. □

Applying (2.2) and (1.5.III) we obtain

7.22 Proposition. *If* X *is completely regular and if* $p : E \to X$ *is any bundle, then the* $C_b(X)$*-module* $\Gamma(p)$ *is reduced if and only if the bundle* $p : E \to X$ *is reduced.* □

7.23 Proposition. *Let* $p : E \to X$ *be a reduced bundle of Banach spaces over a completely regular base space* X*, then*

$$T_{-} : C(X) \to B(\Gamma(p))$$

$$f \to T_f, \quad T_f(a) = f \cdot a$$

is an isometry of Banach algebras.

Proof. Applying (2.10) we obtain

$$\begin{aligned} \|T_f\| &= \sup \{ \|f \cdot \sigma\| : \|\sigma\| \leq 1\} \\ &= \sup_{x \in X} \sup \{|f(x)| \cdot \|\sigma(x)\| : \|\sigma\| \leq 1\} \\ &= \sup_{x \in X} \sup \{|f(x)| \cdot \|\alpha\| : \|\alpha\| \leq 1, \alpha \in E_x\} \\ &= \sup \{|f(x)| : x \in X, p^{-1}(x) \neq 0\} \end{aligned}$$

$$= \|f\|$$

as $\{x \in X : p^{-1}(x) \neq 0\}$ is dense in X. □

7.24 Corollary. *If* E *is a Banach space which is a reduced locally* C(X)*-convex* C(X)*-module for a certain compact space* X*, then the mapping* $f \to T_f : C(X) \to \mathcal{B}(E)$ *is an isometry of Banach algebras.* □

We conclude this section with some examples:

7.25 Let E be a Banach algebra. Recall that the *centroid* $Z_\Omega(E)$ of E is the set of all bounded continuous operators $T : E \to E$ satisfying $a\cdot T(b) = T(a\cdot b) = T(a)\cdot b$ for all $a,b \in E$. If T belongs to the centroid, then $T^\perp = \{a \in E : T(a) = 0\}$ is always a closed ideal.

Now let $p : \mathcal{E} \to X$ be a reduced bundle of Banach algebras. It is easy to verify that for $f \in C_b(X)$ the mapping $\sigma \to f\cdot\sigma : \Gamma(p) \to \Gamma(p)$ belongs to the centroid of $\Gamma(p)$. We shall abreviate this fact by writing $C_b(X) \subset Z_\Omega(\Gamma(p))$.
Conversely, if the Banach algebra E is a reduced $C_b(X)$-module, then $C_b(X) \subset Z_\Omega(E)$ implies that E is a $C_b(X)$-Ω-module.

As in a C^*-algebra every closed ideal is a *-ideal, we can state:

7.26 Corollary. *Let* X *be a compact space and let* E *be a Banach algebra* (C^**-algebra*) *which is at the same time a reduced* C(X)*-module. Then the following statements are equivalent:*

(i) E is locally C(X)-convex and $C(X) \subset Z_\Omega(E)$.

(ii) There is a bundle $p : \mathcal{E} \to X$ of Banach algebras (C^*-algebras) such that E is isometrically isomorphic to $\Gamma(p)$. □

As a matter of fact, for C^*-algebras the inclusion $C(X) \subset Z_\Omega(E)$ <u>implies</u> that E is locally $C(X)$-convex (see section 14).

7.27 Let E be a Banach lattice and let $S,T : E \to E$ be bounded linear operators. We say that $S \leq T$ if $S(a) \leq T(a)$ for all $0 \leq a \in E$. The *center* of E is defined to be the set

$$Z_\Omega(E) := \{T \in \mathcal{B}(E) : -r\cdot Id \leq T \leq r\cdot Id \text{ for some } r \in \mathbb{R}\} .$$

It is known that $Z_\Omega(E)$ is as an ordered vector space and as an algebra over $\mathbb{R}$ isometrically isomorphic to $C(Y)$, where Y is a compact space (see [Wi 71],[FGK 78]). Moreover, for all $T \in Z_\Omega(E)$ and all $a \in E$ we have $|T(a)| = |T|(|a|)$ (this follows immediately from theorem (2.2) in [FGK 78]). Especially, all positive elements $0 \leq T \in Z_\Omega(E)$ are lattice homomorphisms. Hence the equivalences $T(a) = 0$ iff $|T(a)| = 0$ iff $|T|(|a|) = 0$ iff $||T|(a)| = 0$ iff $|T|(a) = 0$ show that ker T is an ideal of E for every $T \in Z_\Omega(E)$.

Now let $p : E \to X$ be a reduced bundle of Banach lattices. Some straightforward arguments show that in this case we have $C_b(X) \subset Z_\Omega(\Gamma(p))$, i.e. the operator $\sigma \to f\cdot\sigma : \Gamma(p) \to \Gamma(p)$ belongs to $Z_\Omega(E)$ for every $f \in C_b(X)$.
Conversely, if E is a Banach lattice which is a reduced $C_b(X)$-module, then the above arguments show that $C_b(X) \subset Z_\Omega(E)$ implies that $f^\perp$ is an ideal of E for every $f \in C_b(X)$. Thus, we have the following analog to (7.26):

7.28 Corollary. *Let X be a compact space and let E be a Banach lattice which also is a reduced C(X)-module. The following statements are equivalent:*

(i) E is locally $C(X)$-convex and $C(X) \subset Z_\Omega(E)$.

(ii) There is a bundle $p : \mathcal{E} \to X$ of Banach lattices such that E is isometrically isomorphic to the Banach lattice $\Gamma(p)$. □

<u>Problem</u>. Is there a general notion of "center" for Ω-spaces in general? If so, can this center be described in the form $C_b(\text{Prim } E)$, where Prim E is a set of "primitive" Ω-ideals of E carrying the hull-kernel topology? (If E is a Banach algebra or a Banach lattice, see [DH 68] and [FGK 78], for a "topological" version of this problem see section 14.)

8. C(X)-submodules of $\Gamma(p)$

Let us suppose that F is a closed submodule of $\Gamma(p)$, where $p : E \to X$ is a bundle of Banach spaces over a compact base space X. Then, of course, F is a locally C(X)-convex C(X)-module, too, and therefore F may be represented as the Banach space of all sections in a bundle $p' : E' \to X$. We shall see that E' may be identified with a certain subset of E and we shall give some characterizations of the subsets of E obtained in this way.

In the beginning of this section we return again to bundles of Ω-spaces with certain families of seminorms:

8.1 Definition. Let $\tau : I \to \mathbb{N}$ be a type and let $p : E \to X$ be a bundle of Ω-spaces of type τ with family of seminorms $(\nu_j)_{j\in J}$. A subset $F \subset E$ is called a Ω-*subbundle* if

(i) $p^{-1}(x) \cap F$ is a (non-empty) Ω-subspace of E_x for every $x \in X$

(ii) Given $\alpha \in F$, $j \in J$ and $\varepsilon > 0$, there is a neighborhood U of $p(\alpha)$ and a section $\sigma \in \Gamma_U(p)$ such that $\sigma(x) \in F$ for all $x \in U$ and such that $\nu_j(\sigma(p(\alpha)) - \alpha) < \varepsilon$.

A subbundle F is called *stalkwise closed*, if $p^{-1}(x) \cap F$ is closed in E_x for every $x \in X$. □

A large part of the following proposition follows immediatly from the definitions:

8.2 Proposition. *Let* $p : E \to X$ *be a bundle of* Ω*-spaces with family of seminorms* $(\nu_j)_{j\in J}$*. If* $F \subset E$ *is a* Ω*-subbundle, then the*

restriction $p_{/F} : F \to X$ *is a bundle of* Ω*-spaces itself having* $(\nu_{j/F})_{j\in J}$ *as a family of seminorms, when we equip* F *with the* Ω*-structure and the topology inherit from* E.
Especially, the restriction of p *to* F *is still open.*

Proof. The only interesting point to prove is the following: If $O \subset F$ is open in F, then O is a union of tubes, i.e. we have to verify axiom (1.5.II).
Let us start with an open set $O \subset F$. Then we may find an open set $O' \subset E$ such that $O = O' \cap F$. Pick any $\alpha \in O$. Then we may find an open neighborhood U_1 of $p(\alpha)$, an index $j \in J$, a real number $\varepsilon > 0$ and a local section $\sigma : U_1 \to E$ such that $\nu_j(\sigma(p(\alpha)) - \alpha) < \varepsilon$ and such that $\{\beta \in E : \nu_j(\beta - \sigma(p(\beta))) < \varepsilon \text{ and } p(\beta) \in U_1\} \subset O'$. Let

$$\delta = \frac{1}{2}(\varepsilon - \nu_j(\sigma(p(\alpha)) - \alpha)).$$

By (8.1(ii)) there is an open neighborhood U_2 of $p(\alpha)$ and a continuous section $\rho : U_2 \to E$ such that $\rho(U_2) \subset F$ and $\nu_j(\rho(p(\alpha)) - \alpha) < \delta$. Obviously we have $\nu_j(\rho(p(\alpha)) - \sigma(p(\alpha))) < \varepsilon - \delta$. Hence there is an open neighborhood $U \subset U_1 \cap U_2$ of $p(\alpha)$ such that $\nu_j(\rho(x) - \sigma(x)) < \varepsilon - \delta$ for all $x \in U$. Moreover, using the triangle inequality we obtain $\nu_j(\beta - \sigma(p(\beta))) < \varepsilon$ whenever $p(\beta) \in U$ and $\nu_j(\beta - \rho(p(\beta))) < \delta$. This yields

$$\alpha \in \{\beta \in F : p(\beta) \in U \text{ and } \nu_j(\beta - \rho(p(\beta))) < \delta\} \subset O. \qquad \square$$

The next result is a trivial remark following from the definitions:

8.3 Proposition. *If* $p : E \to X$ *is a bundle of* Ω*-spaces and if* $F \subset E$ *is an* Ω*-subbundle, then* $\bigcup_{x\in X} \mathrm{cl}(E_x \cap F)$ *is a stalkwise closed* Ω*-subbundle, where* $\mathrm{cl}(E_x \cap F)$ *denotes the closure of* $E_x \cap F$ *in* E_x. $\square$

We now discuss the connection between subbundles of $p : E \to X$ and $C(X)$-submodules of $\Gamma(p)$. First of all, we should remark that a considerations of the Ω-structure makes only sense if $\Gamma(p)$ is a Ω-space and this is only guaranteed if the base space is quasi-compact. This explains the somehow technical postulates in the following proposition:

8.4 Proposition. *Let* $p : E \to X$ *be a bundle of* Ω*-spaces and assume furthermore that* $\Gamma(p)$ *is algebraically an* Ω*-subspace of the cartesian product* $\prod_{x \in X} E_x$

(i) *If* $F \subset E$ *is an* Ω*-subbundle, then* $\Gamma(p_{/F})$ *is an* Ω*-subspace and a* $C_b(X)$*-submodule of* $\Gamma(p)$.

(ii) *Conversely, if* $F \subset \Gamma(p)$ *is an* Ω*-subspace, then* $\bigcup_{x \in X} \varepsilon_x(F)$ *is an* Ω*-subbundle of* E. □

Of course, even if we restrict ourselves to $C(X)$-submodules, there is no reason to believe that this last proposition sets up a one-to-one correspondence between all Ω-subbundles of E and all Ω-subspaces and $C_b(X)$-submodules of $\Gamma(p)$. For instance, all $C_b(X)$-submodules of $\Gamma(p)$ of the form $\Gamma(p_{/F})$, $F \subset E$ a subbundle, are fully additive in the sense of (4.1). However, the following example shows that even for trivial bundles with discrete base space and stalk $\mathbb{R}$ we can find a $C_b(X)$-submodule of $\Gamma(p)$ which is not fully additive:

8.5 Example. Let X be any infinite, non-countable set, equipped with the discrete topology, let $E = X \times \mathbb{R}$ and let $p : E \to X$ be the first projection. Then $\Gamma(p)$ consists of all bounded mappings from X into $\mathbb{R}$. Moreover, if $F \subset \Gamma(p)$ is a fully additive $C_b(X)$-submodule, then

$$F = \{\sigma \in \Gamma(p) : \sigma_{/M} = 0\} \quad \text{where} \quad M = \bigcap_{\sigma \in F} \sigma^{-1}(0).$$

Indeed, the inclusion $F \subset \{\sigma \in \Gamma(p) : \sigma_{/M} = 0\}$ holds trivially. Conversely, suppose that $\sigma_{/M} = 0$ and let $x \in X \setminus M$. Then there is an element $\tau \in F$ such that $\tau(x) \neq 0$. Multiplying τ with the continuous function $\frac{\sigma(x)}{\tau(x)} \cdot \chi_x$, where $\chi_x(x) = 1$ and $\chi_x(y) = 0$ for $x \neq y$, we obtain an element $\tau_x \in F$ such that $\tau_x(x) = \sigma(x)$ and $\tau_x(y) = 0$ for $x \neq y$. Clearly, the family $(\tau_x)_{x \in X \setminus M}$ is locally finite and $\sigma = \sum_{x \in X \setminus M} \tau_x$. This proves that $\sigma \in F$, as F is fully additive.

Now let $F_c := \{\sigma \in \Gamma(p) : \sigma^{-1}(\mathbb{R} \setminus \{0\})$ is countable$\}$. Then F_c is a closed $C(X)$-submodule of $\Gamma(p)$. As $\bigcap_{\sigma \in F_c} \sigma^{-1}(0) = \emptyset$ and as $F_c \neq \Gamma(p)$, F_c is not fully additive.

8.6 Theorem. *Let* $p : E \to X$ *be a bundle of* Ω*-spaces and assume that one of the following two conditions is satisfied:*

(i) The base space X is compact.

or

(ii) The base space X is paracompact, $p : E \to X$ is a bundle of Banach spaces and $\Gamma(p)$ is an Ω-subspace of $\prod_{x \in X} E_x$.

Then the following statements are true:

(a) If $F \subset E$ *is a stalkwise closed* Ω*-subbundle, then* $\Gamma(p_{/F})$ *is a fully additive closed* Ω*-submodule of* $\Gamma(p)$.

(b) If $F \subset \Gamma(p)$ *is a fully additive closed* Ω*-submodule of* $\Gamma(p)$, *then* $\bigcup_{x \in X} cl(\varepsilon_x(F)) =: E_F$ *is a stalkwise closed* Ω*-subbundle of* E.

Moreover, the mapping $F \to \Gamma(p_{/F})$ *is a bijection between the set of all stalkwise closed* Ω*-subbundles and the set of all fully additive closed* Ω*-submodules of* $\Gamma(p)$. *The inverse of this mapping is given by* $F \to E_F$.

Proof. From the Stone-Weierstraß theorem (4.2) (or (4.3) resp.) we conclude that F is dense in $\Gamma(p_{/E_F})$. As F is closed in $\Gamma(p)$, we obtain equality.
Conversely, by the definitions we have $\varepsilon_x(\Gamma(p_{/F})) \subset E_x \cap F$, and the smaller set is dense in the larger one. As by assumption F was stalkwise closed, we obtain equality in this case, too. □

8.7 Remarks. (i) If X is compact, then we do not have to postulate that F is fully additive: In this case, every C(X)-submodule of $\Gamma(p)$ is fully additive.
(ii) If all stalks of the bundle $p : E \to X$ are complete and if the bundle has a countable family of seminorms $(\nu_j)_{j\in J}$, then $\varepsilon_x(F)$ is automatically closed in E_x, provided that F is a closed and fully additive $C_b(X)$-submodule of $\Gamma(p)$. Hence, under these conditions, we may set $E_F = \bigcup_{x\in X} \varepsilon_x(F)$.
(Indeed, from (8.6) we know that $F = \Gamma(p_{/E_F})$, where $E_F = \bigcup_{x\in X} cl(\varepsilon_x F)$. As the bundle $p_{/E_F} : E_F \to X$ is full by (2.9), for a given $\alpha \in cl(\varepsilon_x(F))$ we may find a $\sigma \in F$ such that $\sigma(p(\alpha)) = \alpha$, i.e. $\varepsilon_x(F) = cl(\varepsilon_x(F))$.)

As in section 3 we may virtually "weaken" the notion of subbundles in certain situations. We shall do this in the following proposition, which is an immediate consequence of (3.2):

8.8 Proposition. *Let* $p : E \to X$ *be a bundle of* Ω*-spaces and suppose that one of the following two properties are satisfied:*

(a) $p : E \to X$ is a bundle of Banach spaces and X is locally paracompact.

(b) $p : E \to X$ has a locally countable family of seminorms $(\nu_j)_{j\in J}$, all stalks are semicomplete and the base space X is locally compact.

Then $F \subset E$ *is a stalkwise closed* Ω*-subbundle if and only if*

(i) $F \cap E_x$ *is a closed* Ω*-subspace of* E_x *for every* $x \in X$.

(ii) *The restriction* $p_{/F} : F \to X$ *is still open.* □

Thus, if we restrict ourselves to bundles of Banach spaces with a compact base space X, we are lead to a study of those "distributions" of closed subspaces $(F_x)_{x \in X}$ of the stalks such that the restriction of the projection $p : E \to X$ to $\bigcup_{x \in X} F_x$ is still open. We shall return to a further discussion of this topic in section 15.

9. Quotients of bundles and C(X)-modules

In the same way we can form quotients of a single topological vector space, we may form quotients of bundles of vector spaces. As one might expect, these quotients a closely related to quotient maps between the corresponding C(X)-modules of sections.

Let $p : E \to X$ be a fixed bundle of Ω-spaces with seminorms $(\nu_j)_{j \in J}$ and let $F \subset E$ be a stalkwise closed subbundle. This time we do not require that $F \cap E_x$ is an Ω-subspace of E_x, but we postulate that $F \cap E_x$ is an Ω-ideal of E_x. Let us agree to call such a subbundle a *stalkwise Ω-ideal.*

A straightforeward proof shows:

9.1 Proposition. *If the subbundle $F \subset E$ is a stalkwise Ω-ideal and if $\Gamma(p)$ is an Ω-subspace of the cartesian product of the stalks, then $\Gamma(p_{/F})$ is an Ω-ideal of $\Gamma(p)$.* □

Hence we may form the quotient $\Gamma(p)/\Gamma(p_{/F})$. It is fairly easy to see that $\Gamma(p)/\Gamma(p_{/F})$ is a topological Ω-space and a locally C(X)-convex C(X)-module if we define

$$f\cdot(\sigma + \Gamma(p_{/F})) := f\cdot\sigma + \Gamma(p_{/F}) \quad \text{for all } \sigma \in \Gamma(p),\ f \in C_b(X).$$

It is less obvious to see that $\Gamma(p)/\Gamma(p_{/F})$ is even a $C_b(X)$-Ω-module in the sense of (7.1), and we shall for the moment accept this fact without proof. Hence, applying (7.16) and (7.17), we are led to the conclusion that, at least for compact base spaces X, the quotient $\Gamma(p)/\Gamma(p_{/F})$ may be represented as the Ω-space of all sections in a suitable bundle of Ω-spaces $q : E' \to X$. As this idea works only for compact base spaces and does not tell very much about the relationship between the bundles E, F and E', we shall turn our attention to another aspect which will yield the above facts automatically.

9.2 Again, let $p : E \to X$ be a bundle of Ω-spaces and let $F \subseteq E$ be a subbundle which is stalkwise a closed Ω-ideal. We define an equivalence relation Θ_F on E by setting

$$(\alpha, \beta) \in \Theta_F \text{ iff } p(\alpha) = p(\beta) \text{ and } \alpha - \beta \in F.$$

Let $E/F := E/\Theta_F$ and let $\pi_F : E \to E/F$ be the quotient map. We equip E/F with the quotient topology. There is more structure we can add to E/F:
First of all, note that $\Theta_F \subset \ker p$. Hence there is a mapping $p_F : E/F \to X$ such that $p = p_F \circ \pi_F$, i.e. the diagram

$$\begin{array}{ccc} E & \overset{\pi_F}{\to} & E/F \\ p \downarrow & & \downarrow p_F \\ X & \underset{id_X}{\to} & X \end{array}$$

commutes. By definition of the topology on E/F, the mapping p_F is continuous.
As $p_F^{-1}(x) = E_x/(F \cap E_x)$ for every $x \in X$, the stalks of $p_F : E/F \to X$ carry an unique Ω-space structure so, that $\pi_F : E \to E/F$ induces stalkwise a homomorphism of Ω-spaces (see (6.3)). Hence $p_F : E/F \to X$ is a fibred Ω-space.
Finally, we define a family $(\nu_j^F)_{j \in J}$ of seminorms on E/F by

$$\nu_j^F(\alpha') := \inf \{\nu_j(\beta) : \beta \in \pi_F^{-1}(\alpha')\} \quad \text{for all } \alpha' \in E/F,$$

i.e. ν_j^F is stalkwise the quotient seminorm of ν_j modulo F.

Of course, we now wish to show that $p_F : E/F \to X$ is a bundle of

Ω-spaces with seminorms $(\nu_j^F)_{j\in J}$ and that $\Gamma(p_F)$ contains $\Gamma(p)/\Gamma(p_{/F})$ as an Ω-subspace (which then will yield a proof for the fact that $\Gamma(p)/\Gamma(p_{/F})$ is an $C_b(X)$-Ω-module).

We shall split the proof into a number a small steps:

9.3 If $U \subset X$ is open and if $\sigma : U \to E$ is a local section, then the mapping $T_\sigma : \alpha \to \alpha + \sigma(p(\alpha)) : p^{-1}(U) \to p^{-1}(U)$ is a homeomorphism. (This observation is already contained in the proof of (5.7)).

9.4 The mapping $\pi_F : E \to E/F$ is open.

(Let $O \subset E$ be open. We have to show that $\pi_F^{-1}(\pi_F(O))$ is open. We claim that

$$\pi_F^{-1}(\pi_F(O)) = \{\alpha\colon p(\alpha) = p(\beta) \text{ and } \alpha - \beta \in F \text{ for some } \beta \in O\}$$
$$= \{\alpha\colon \text{there is an open neighborhood } V \subset p(O) \text{ of } p(\alpha) \text{ and a local section } \sigma : V \to F \text{ such that } \alpha \in T_\sigma(O \cap p^{-1}(V))\}.$$

Indeed, if α is contained in the latter set, then $\alpha = \beta + \sigma(p(\alpha))$ where $\sigma \in \Gamma_V(p_{/F})$. As $\beta \in O$ and as $\alpha - \beta = \sigma(p(\alpha)) \in F$, we obtain $\alpha \in \pi_F^{-1}(\pi_F(O))$.

Conversely, let $\alpha \in \pi_F^{-1}(\pi_F(O))$. We have to find an open neighborhood $V \subset p(O)$ of $p(\alpha)$ and a local section $\sigma : V \to E$ such that $\alpha = \sigma(p(\alpha)) + \beta$ for a certain $\beta \in O$. Firstly, choose a $\beta' \in O$ such that $p(\alpha) = p(\beta')$ and $\alpha - \beta' \in F$. Then select an element $j \in J$ and an $\varepsilon > 0$ such that $\{\gamma : p(\gamma) = p(\alpha) \text{ and } \nu_j(\gamma - \beta') < \varepsilon\} \subset O$. By the definition of subbundles there is an open neighborhood $V \subset p(O)$ of $p(\alpha)$ and a local section $\sigma : V \to F$ such that $\nu_j(\alpha - \beta' - \sigma(p(\alpha))) < \varepsilon$. Let $\beta = \alpha - \sigma(p(\alpha))$. Then $\beta \in O$, as desired.

Now we conclude that

$$\pi_F^{-1}(\pi_F(O)) = \cup\,\{T_\sigma(O \cap p^{-1}(V)) : V \subset p(O) \text{ open}, \sigma \in \Gamma_V(p_{/F})\}$$

and this set is open by (9.3).)

9.5 The mappings

$$\text{add} : (E/F) \vee (E/F) \to E/F$$

$$\text{scal} : \mathbb{K} \times (E/F) \to E/F$$

$$0 : X \to E/F$$

as well as the additional mappings

$$f_i : \bigvee^{\tau(i)} (E/F) \to E/F \qquad i \in I$$

are continuous.

(As all the proofs are similar, we show only the continuity of the mappings $f_i : \bigvee^{\tau(i)} (E/F) \to E/F$.

Firstly, note that the mapping $\pi_F : E \to E/F$ induces a mapping

$$\bigvee^{\tau(i)} \pi_F : \bigvee^{\tau(i)} E \to \bigvee^{\tau(i)} (E/F)$$

$$(\alpha_1, \ldots, \alpha_{\tau(i)}) \to (\pi_F(\alpha_1), \ldots, \pi_F(\alpha_{\tau(i)})).$$

By the definition of the topologies on $\bigvee^{\tau(i)} E$ and $\bigvee^{\tau(i)} (E/F)$ resp., which is essentially the product topology, and by (.4), the mapping $\bigvee^{\tau(i)} \pi_F$ is surjective, continuous and open, whence a quotient map. As the mapping $f_i : \bigvee^{\tau(i)} E \to E$ is continuous, the assertion now follows from the commutativity of the diagram

$$\begin{array}{ccc} \bigvee^{\tau(i)} E & \xrightarrow{\bigvee^{\tau(i)} \pi_F} & \bigvee^{\tau(i)} (E/F) \\ f_i \downarrow & & \downarrow f_i \\ E & \xrightarrow[\pi_F]{} & E/F \end{array}$$)

9.6 Given $\alpha' \in E/F$, $j \in J$ and $\varepsilon > 0$, there is an open neighborhood U of $p_F(\alpha')$ and a continuous section $\sigma' : U \to E/F$ such that $\nu_j^F(\sigma'(p_F(\alpha') - \alpha')) < \varepsilon$.

(Let $\alpha \in \pi_F^{-1}(\alpha')$. Then we may find an open neighborhood U of $p(\alpha) = p_F(\alpha')$ and a continuous section $\sigma \in \Gamma_U(p)$ such that $\nu_j(\sigma(p(\alpha)) - \alpha) < \varepsilon$. Define $\sigma' := \pi_F \circ \sigma$. Then σ' has the desired properties.)

As F is stalkwise closed, (1.5.IV) and (1.6(viii)) imply

9.7 If $\alpha \in E/F$, then $\alpha = 0$ if and only if $\nu_j^F(\alpha) = 0$ for all $j \in J$.

It remains to check axiom (1.5.II), i.e. we have to show that the tubes form a base for the topology on E/F:

9.8 If $O' \subset E/F$ is open and if $\alpha' \in O'$, then we can find an open set $U \subset X$, a continuous section $\sigma' : U \to E/F$, a $j \in J$ and a real number $\varepsilon > 0$ such that the tube $T(U,\sigma',\varepsilon,j)$ is open and satisfies

$$\alpha' \in \{\beta' \in E/F : p\ (\beta') \in U \text{ and } \nu_j^F(\sigma'(p_F(\beta') - \beta') < \varepsilon\} \subset O'.$$

(Let $\alpha \in \pi_F^{-1}(\alpha') \subset \pi_F^{-1}(O') =: O$. Then there is an open set $U \subset X$, a section $\sigma \in \Gamma_U(p)$, a $j \in J$ and an $\varepsilon > 0$ such that

$$\alpha \in \{\beta \in E : p(\beta) \in U \text{ and } \nu_j(\sigma(p(\beta)) - \beta) < \varepsilon\} \subset O.$$

Let $\sigma' := \pi_F \circ \sigma$. As usual, we abbreviate

$$T(U,\sigma,\varepsilon,j) = \{\beta \in E : p(\beta) \in U \text{ and } \nu_j(\sigma(p(\beta)) - \beta) < \varepsilon\} \quad \text{and}$$
$$T(U,\sigma',\varepsilon,j) = \{\beta' \in E/F : p_F(\beta') \in U \text{ and } \nu_j^F(\sigma'(p_F(\beta')) - \beta') < \varepsilon\}$$

The proof of (9.8) will be complete if we can show that

$$\pi_F(T(U,\sigma,\varepsilon,j)) = T(U,\sigma',\varepsilon,j)$$

as then we can conclude that $\alpha' \in T(U,\sigma',\varepsilon,j) \subset O'$. Moreover, the set $T(U,\sigma',\varepsilon,j)$ will be open, as the mapping π_F is open.
The inclusion $\pi_F(T(U,\sigma,\varepsilon,j)) \subset T(U,\sigma',\varepsilon,j)$ is easy to see, as by

definition we have $\nu_j^F(\sigma'(p_F(\pi_F(\beta))) - \pi_F(\beta)) = \nu_j^F(\pi_F(\sigma(p(\beta)) - \beta)) \le$
$\le \nu_j(\sigma(p(\beta)) - \beta)$.
Conversely, let $\beta' \in T(U,\sigma',\varepsilon,j)$. Then we know that $\nu_j^F(\sigma'(p_F(\beta')) -$
$- \beta') < \varepsilon$. Hence there is an element $\gamma \in \pi_F^{-1}(\sigma'(p_F(\beta')) - \beta')$ such that $\nu_j(\gamma) < \varepsilon$. Define $\beta = \sigma(p(\gamma)) - \gamma$. Then β belongs to $T(U,\sigma,\varepsilon,j)$ and $\pi_F(\beta) = \beta'$, i.e. $\beta' \in \pi_F(T(U,\sigma,\varepsilon,j))$.)

9.9 The mapping

$$\Pi_F : \Gamma(p) \to \Gamma(p_F)$$
$$\sigma \to \pi_F \circ \sigma$$

is a continuous homomorphism of $C_b(X)$-modules and Ω-spaces. Moreover, $\ker \Pi_F = \Gamma(p_{/F})$.
(We only have to prove the continuity of Π_F. But this follows immediately from $\vartheta_j^F(\Pi_F(\sigma)) \le \vartheta_j(\sigma)$.)

9.10 If X is compact <u>or</u> if X is paracompact and if $p : E \to X$ is a bundle of normed spaces, then we have

$$\vartheta_j^F(\Pi_F(\sigma)) = \inf \{\vartheta_j(\sigma + \rho) : \rho \in \Gamma(p_{/F})\}.$$

Especially, in both cases, the mapping Π_F is open onto its image.
(It is easy to check that we always have $\vartheta_j^F(\Pi_F(\sigma)) \le \inf \{\vartheta_j(\sigma + \rho) :$
$\rho \in \Gamma(p_{/F})\}$.
Conversely, let $M = \vartheta_j^F(\Pi_F(\sigma))$ and let $\varepsilon > 0$. Then for every $x \in X$ we have $\nu_j^F(\Pi_F(\sigma)(x)) = \nu_j^F(\pi_F(\sigma(x))) < M + \varepsilon$. Hence there is a certain $\alpha \in F \cap E_x$ such that $\nu_j(\sigma(x) + \alpha) < M + \varepsilon$. Let

$$\delta := M + \varepsilon - \nu_j(\sigma(x) + \alpha).$$

By the definition of subbundles, there is a local section $\rho_x : U \to F$ such that $\nu_j(\alpha - \rho_x(p(\alpha))) < \delta$; using (2.2) we may assume that

$\rho_x \in \Gamma(p_{/F})$ is a global section. As $\nu_j(\sigma(x) + \rho_x(x)) < M + \varepsilon$, there is an open neighborhood U_x of x such that $\nu_j(\sigma(y) + \rho_x(y)) < M + \varepsilon$ for all $y \in U_x$. We now follow the path we have walked several times before: By passing to a refinement if necessary, we may assume that the covering $(U_x)_{x \in X}$ is locally finite. Take a partition of unity $(f_x)_{x \in X}$ subordinate to $(U_x)_{x \in X}$. Now define

$$\rho := \sum_{x \in X} f_x \cdot \rho_x$$

As the family $(f_x \cdot \rho_x)_{x \in X}$ is locally finite, ρ maps X into F and is a continuous selection. Moreover, ρ is bounded: This follows trivially in the case where the base space X is compact. If $p : E \to X$ is a bundle of normed spaces, then the family of seminorms consists of one element only, namely ν_j. Hence we only have to show that the mapping $x \to \nu_j(\rho(x)) : X \to \mathbb{R}$ is bounded. But this follows easily from the triangle inequality and the following

$$\begin{aligned} \nu_j(\sigma(y) + \rho(y)) &\le \sum_{x \in X} f_x(y)\,\nu_j(\sigma(y) + \rho_x(y)) \\ &\le \sum_{x \in X} f_x(y) \cdot (M + \varepsilon) \\ &= M + \varepsilon, \end{aligned}$$

as we have either $y \in U_x$ and then $\nu_j(\sigma(y) + \rho_x(y)) \le M + \varepsilon$ or we have $y \notin U_x$ in which case $f_x(y) = 0$.

Hence in both cases, ρ will be a continuous section $\rho : X \to F$. Moreover, the above argument shows that $\vartheta_j(\sigma + \rho) \le M + \varepsilon$. This yields the inequality $\inf \{\vartheta_j(\sigma + \rho) : \rho \in \Gamma(p_{/F})\} \le \vartheta_j^F(\Pi_F(\sigma)) + \varepsilon$. As $\varepsilon > 0$ was arbitrary, the proof is complete.)

From the Stone-Weierstraß theorem (4.2) we conclude:

9.11 Under the hypothesis of (9.10), the image of Π_F is dense in

$\Gamma(p_F)$.

9.12 If the bundle $p : E \to X$ satisfies the assumptions of (9.10) and if in addition all stalks of the bundle are complete and if the family of seminorms is countable, then the mapping $\Pi_F : \Gamma(p) \to \Gamma(p_F)$ is surjective.

(We already know that $\Pi_F(\Gamma(p))$ is dense in $\Gamma(p_F)$ and that $\Pi_F(\Gamma(p))$ is topologically and algebraically isomorphic to $\Gamma(p)/\Gamma(p_{/F})$. Moreover, by assumption and (1.10), $\Gamma(p)$ is complete and metric. As the quotient of a complete metric space is again complete, $\Gamma(p)/\Gamma(p_{/F})$ and therefore $\Pi_F(\Gamma(p))$ are complete. This yields $\Pi_F(\Gamma(p)) = \Gamma(p_F)$.)

We collect all these partial results in a theorem:

9.13 Theorem. *Let* $p : E \to X$ *be a bundle of* Ω*-spaces with seminorms* $(\nu_j)_{j\in J}$. *Moreover, let* $F \subset E$ *be a stalkwise closed subbundle, which is stalkwise an* Ω*-ideal. Then* $p_F : E/F \to X$ *is a bundle of* Ω*-spaces, where* E/F *carries the quotient topology and the quotient structure of* Ω*-spaces.*

If $\pi_F : E \to E/F$ *is the quotient map, then*

$$\Pi_F : \Gamma(p) \to \Gamma(p_F)$$
$$\sigma \to \pi_F \circ \sigma$$

is a continuous homomorphism with kernel $\Gamma(p_{/F})$.

If X *is compact, then* Π_F *is open onto its image and the image of* Π_F *is dense.*

Finally, if all stalks of E *are complete and if the family of seminorms is countable, then* Π_F *is surjective.* □

9.14 Theorem. *Let* $p : E \to X$ *be a bundle of Banach spaces over a paracompact base space and let* $F \subset E$ *be a stalkwise closed subbundle. If we equip the stalkwise quotient* E/F *with the quotient topology, the quotient structure and the quotient norm, then we obtain a bundle of Banach spaces* $p_F : E/F \to X$. *Moreover, the quotient space* $\Gamma(p)/\Gamma(p_{/F})$ *is canonically isomorphic and isometric to* $\Gamma(p_F)$. □

10. Morphisms between bundles.

Having discussed subobjects and quotients, we should also make some remarks on morphisms between bundles in general. As everybody would expect by now, these morphisms will be closely related with homomorphisms between the corresponding spaces of sections.

10.1 Definition. (i) Let E and F be Ω-spaces which are at the same time $C_b(X)$-modules. A linear map $T : E \to F$ is called a $C_b(X)$-*Ω-morphism*, if T is a homomorphism of Ω-spaces also preserving the $C_b(X)$-module structure.

(ii) Let $p : E \to X$ and $q : F \to X$ be bundles of Ω-spaces of the same type τ and with seminorms $(\nu_j)_{j\in J}$ and $(\mu_k)_{k\in K}$ resp. A mapping $\lambda : E \to F$ is called a *morphism of Ω-bundles*, if

a) λ is continuous.

b) $p = q\circ\lambda$, i.e. λ preserves stalks.

c) $\lambda_{/p^{-1}(x)} : p^{-1}(x) \to q^{-1}(x)$ is a homomorphism of Ω-spaces.

d) For every $k \in K$ there are elements $j \in J$ and $0 < M \in \mathbb{R}$ such that $\nu_j(\alpha) \le M$ implies $\mu_k(\lambda(\alpha)) \le 1$. □

The property d) in some sense says that the family $(\lambda_{/p^{-1}(x)})_{x\in X}$ has to be "equicontinuous". We shall illustrate this statement in example (10.20).

From the point of view of "equicontinuity" it is not suprising that property d) holds automatically if X is compact:

10.2 Proposition. *If $p : E \to X$ and $q : F \to X$ are bundles with compact base space X and seminorms $(\nu_j)_{j\in J}$ and $(\mu_k)_{k\in K}$ resp., and if $\lambda : E \to F$ is any map, then the properties* (a), (b) *and* (c) *of* (10.1) *imply property* (d).

Proof. Let $k \in K$ be any index. Then $\{\beta \in F : \mu_k(\beta) < 1\}$ is open in F. As λ is continuous, the set $O := \lambda^{-1}(\{\beta \in F : \mu_k(\beta) < 1\})$ is open in E and contains the 0 of E_x for every $x \in X$. Hence, for each $x \in X$ we may find an open neighborhood U_x of x, an $\varepsilon_x > 0$ and an element $j_x \in J$ such that $\{\alpha \in E : p(\alpha) \in U_x$ and $\nu_{j_x}(\alpha) < \varepsilon_x\} \subset O$. Now the compactness of X yields finitely many points $x_1, \ldots, x_n \in X$ such that $U_{x_1} \cup \ldots \cup U_{x_n} = X$. Let $M = \min\{\varepsilon_{x_1}, \ldots, \varepsilon_{x_n}\}$ and let $j \in J$ be any element such that $j_{x_i} \leq j$ for all $1 \leq i \leq n$. Then we obtain $\{\alpha \in E : \nu_j(\alpha) < M\} \subset \lambda^{-1}(O)$. This of course implies $\mu_k(\lambda(\alpha)) \leq 1$ whenever $\nu_j(\alpha) \leq M$. □

We now enter the discussion of the connection between $C_b(X)$-Ω-module morphisms and morphisms between bundles of Ω-spaces. The next proposition is straightforward:

10.3 Proposition. *Let $p : E \to X$ and $q : F \to X$ be bundles of Ω-spaces and assume that $\Gamma(p)$ and $\Gamma(q)$ are Ω-subspaces of the resp. cartesian product of their stalks. If $\lambda : E \to F$ is a morphism between bundles of Ω-spaces, then*

$$T_\lambda : \Gamma(p) \to \Gamma(q) \qquad \textit{defined by}$$

$$T_\lambda(\sigma)(x) = \lambda(\sigma(x)) \qquad \sigma \in \Gamma(p),\ x \in X$$

is a continuous $C_b(X)$-Ω-module homomorphism. □

The following example shows that a converse of (10.3) does not always hold:

10.4 Example. Let c_o be the Banach space of all convergent sequences in $\mathbb{K}$ with limit 0, equipped with the supremum norm. As we know from (5.15), we may identify c_o with the space $\Gamma(p)$, where

$p : E \to \mathbb{N}$ is a bundle of Banach spaces whose stalks are all isomorphic to $\mathbb{K}$ and where $\mathbb{N}$ carries the cofinal topology. As every continuous $f : \mathbb{N} \to \mathbb{K}$ is constant, every bounded linear operator $T : c_o \to c_o$ is a $C_b(\mathbb{N})$ - module homomorphism. Especially, the shift $S : c_o \to c_o$, $S((u_n)_{n\in N}) = S((u_{n+1})_{n\in N})$ is a $C(\mathbb{N})$-module homomorphism, which is not induced by a bundle morphism.

This example shows that the operator T has at least to leave the subspaces N_x invariant in order to be induced by a bundle morphism:

10.5 Proposition. *Let* $p : E \to X$ *and* $q : F \to X$ *be bundles of* Ω*-spaces and assume that* $\Gamma(p)$ *and* $\Gamma(q)$ *are* Ω*-subspaces of the direct product of their stalks.*
Moreover, assume that $p : E \to X$ *is a full bundle and that for every* $x \in X$ *the evaluation map* $\varepsilon_x : \Gamma(p) \to p^{-1}(x)$ *is a quotient map in the sense that* $\nu_j(\alpha) = \inf\{\vartheta_j(\sigma) : \sigma \in \Gamma(p),\ \sigma(p(\alpha)) = \alpha\}$*, where* ν_j *denotes any of the seminorms belonging to the bundle* $p : E \to X$.
If $T : \Gamma(p) \to \Gamma(q)$ *is a continuous morphism between* Ω*-spaces such that*

$$T(\{\sigma \in \Gamma(p) : \sigma(x) = 0\}) \subset \{\rho \in \Gamma(q) : \rho(x) = 0\},$$

then there is a morphism of bundles of Ω*-spaces* $\lambda : E \to F$ *such that* $T = T_\lambda$.
Especially, T *is a* $C_b(X)$*-module homomorphism.*

Proof. As $\Gamma(p)$ is a full bundle, the evaluation map $\varepsilon_x : \Gamma(p) \to p^{-1}(x)$ is surjective. If we denote the evaluation map $\Gamma(q) \to q^{-1}(x)$ with ε_x, too (and hope that this will confuse nobody), then the assumption that T maps $\{\sigma \in \Gamma(p) : \sigma(x) = 0\}$ into $\{\rho \in \Gamma(q) : \rho(x) = 0\}$ is equivalent to $\ker \varepsilon_x \subset \ker(\varepsilon_x \circ T)$. Applying (6.2), (6.3) and (6.4), we find an Ω-morphism $\lambda_x : p^{-1}(x) \to q^{-1}(x)$ such that $\lambda_x \circ \varepsilon_x = \varepsilon_x \circ T$.

$$\begin{array}{ccc} \Gamma(p) & \xrightarrow{T} & \Gamma(q) \\ \varepsilon_x \downarrow & & \downarrow \varepsilon_x \\ p^{-1}(x) & \xrightarrow[\lambda_x]{} & q^{-1}(x) \end{array}$$

We now define $\lambda : E \to F$ by $\lambda(\alpha) := \lambda_{p(\alpha)}(\alpha)$. Then, by construction, (b) and (c) of (10.1(ii)) are satisfied. Clearly, we have $T(\sigma)(x) = (\varepsilon_x \circ T)(\sigma) = (\lambda_x \circ \varepsilon_x)(\sigma) = \lambda_x(\sigma(x)) = \lambda(\sigma(x))$, i.e. $T = T_\lambda$. It remains to check that λ has the properties (a) and (d) of (10.1(ii)).

Let $(\nu_j)_{j \in J}$ and $(\mu_k)_{k \in K}$ be the seminorms of $p : E \to X$ and $q : F \to X$, respectively.
As $T : \Gamma(p) \to \Gamma(q)$ is continuous, for every $k \in K$ there is an $M > 0$ and an $j \in J$ such that $\hat{\nu}_j(\sigma) < M$ implies $\hat{\mu}_k(T(\sigma)) \leq 1$. Now let $\alpha \in E$ be such that $\nu_j(\alpha) < M$. As the evaluation map $\varepsilon_x : \Gamma(p) \to p^{-1}(p(\alpha))$ is a quotient map in the sense that $\nu_j(\beta) = \inf \{\hat{\nu}_j(\sigma) : \sigma(p(\beta)) = \beta\}$, we can find a section $\sigma \in \Gamma(p)$ such that $\hat{\nu}_j(\sigma) < M$ and $\sigma(p(\alpha)) = \alpha$. By the choice of j and M this implies $\hat{\mu}_k(T(\sigma)) \leq 1$ and especially $\mu_k(\lambda(\alpha)) = \mu_k(T(\sigma)(p(\alpha))) \leq 1$. This shows (d) of (10.1.(ii)).

Finally, we show that $\lambda : E \to F$ is continuous: Let $\alpha \in E$ and let $\mathcal{O}$ be any open neighborhood of $\lambda(\alpha)$. As the bundle $p : E \to X$ is full, there is a section $\sigma \in \Gamma(p)$ such that $\sigma(p(\alpha)) = \alpha$. Hence $T(\sigma)$ is a continuous section of $q : F \to X$ passing through $\lambda(\alpha)$. Therefore, by (1.6(vii)), there is an open neighborhood U of $p(\alpha) = q(\lambda(\alpha))$, an $\varepsilon > 0$ and an $k \in K$ such that

$$\{\beta \in F : q(\beta) \in U \text{ and } \mu_k(T(\sigma)(p(\beta)) - \beta) < \varepsilon\} \subset \mathcal{O}.$$

As in the proof of (d), we pick $M > 0$ and $j \in J$ such that $\hat{\nu}_j(\rho) < M$

implies $\hat{\mu}_k(T(\rho)) < 1$. We claim that the open set

$$\{\beta \in E : p(\beta) \in U \text{ and } \nu_j(\beta - \sigma(p(\beta))) < M\cdot\varepsilon\}$$

is contained in $\lambda^{-1}(O)$ and thus λ is continuous at α:
Assume that $p(\beta) \in U$ and $\nu_j(\beta - \sigma(p(\beta))) < M\cdot\varepsilon$. By our assumptions, we can find a section $\tau \in \Gamma(p)$ such that $\tau(p(\beta)) = \beta - \sigma(p(\beta))$ and $\hat{\nu}_j(\tau) < M\cdot\varepsilon$. This implies $\hat{\mu}_k(T(\tau + \sigma) - T(\sigma)) = \hat{\mu}_k(T(\tau)) < 1\cdot\varepsilon = \varepsilon$. Especially, evaluating this inequality at $p(\beta)$, we obtain $\mu_k(\lambda(\beta) - T(\sigma)(p(\beta))) < \varepsilon$, i.e. $\lambda(\alpha) \subset O$. □

Let us recall from (2.5) that the evaluation maps $\varepsilon_x : \Gamma(p) \to p^{-1}(p(x))$ are automatically quotient maps in the sense of (10.5), if the base space X is completely regular and if the bundle $p : E \to X$ is full. Moreover, we may apply (2.9) to obtain the fullness of $p : E \to X$ in certain cases. In these cases we would only have to check whether $T(\{\sigma \in \Gamma(p) : \sigma(x) = 0\}) \subset \{\sigma \in \Gamma(q) : \sigma(x) = 0\}$. But if X is completely regular and if T is a C(X)-module homomorphism, this is always true:

10.6 Proposition. *Let* $p : E \to X$ *and* $q : F \to X$ *be bundles with a completely regular base space* X *and let* $T : \Gamma(p) \to \Gamma(q)$ *be a continuous* $C_b(X)$*-module homomorphism. If* $\sigma \in \Gamma(p)$ *is a section, then* $\sigma(x) = 0$ *implies* $T(\sigma)(x) = 0$.

Proof. By (2.11) it is enough to consider sections of the form $f\cdot\sigma$, where $f \in C_b(X)$ and $f(x) = 0$. But in this case we have obviuously $T(f\cdot\sigma)(x) = (f\cdot T(\sigma))(x) = f(x)\cdot(T(\sigma)(x)) = 0$. □

Hence, for bundles with a completely regular base space, we have the following corollary:

10.7 Corollary. *Let* $p : E \to X$ *and* $q : F \to X$ *be bundles of* Ω*-spaces having a completely regular base space* X. *Suppose in addition that* $\Gamma(p)$ *and* $\Gamma(q)$ *are* Ω*-spaces (which holds automatically, if* X *is compact). If* $T : \Gamma(p) \to \Gamma(q)$ *is a continuous* $C_b(X)$*-*Ω*-module homomorphism, then each of the following conditions implies that* T *is of the form* T_{λ_T}, *where* $\lambda_T : E \to F$ *the morphism of* Ω*-bundles constructed in* (10.5)*:*

(1) The bundle $p : E \to X$ is full.

(2) The bundle $p : E \to X$ has a locally countable family of seminorms and all stalks $(p^{-1}(x))_{x \in X}$ are complete.

Moreover, in these cases the assignment $\lambda \to T_\lambda$ *is a bijection between all* Ω*-bundle morphisms from* E *into* F *and all continuous* $C_b(X)$*-*Ω*-module homomorphisms from* $\Gamma(p)$ *into* $\Gamma(q)$ *with inverse* $T \to \lambda_T$.

Proof. It remains to show that $\lambda_{T_\lambda} = \lambda$: Let $\alpha \in E$. As in both cases (1) and (2) the bundle $p : E \to X$ is full, there is a section $\sigma \in \Gamma(p)$ such that $\sigma(p(\alpha)) = \alpha$. Then an easy calculations gives $\lambda_{(T_\lambda)}(\alpha) = \lambda_{(T_\lambda)}(\sigma(p(\alpha))) = T_\lambda(\sigma)(p(\alpha)) = (\lambda \circ \sigma)(p(\alpha)) = \lambda(\alpha)$, i.e. $\lambda_{(T_\lambda)} = \lambda$. □

Applying (10.7) to isomorphisms $T : \Gamma(p) \to \Gamma(q)$ only, we get an answer to the question to what extent the space of all sections determines the bundle up to isomorphy:

10.8 Definition. Two bundles $p : E \to X$ and $q : F \to X$ of Ω-spaces are called *isomorphic*, if there is a bijection $\lambda : E \to F$ such that λ and λ^{-1} are morphisms of Ω-bundles. □

Clearly, every isomorphism of bundles if a homeomorphism. Conversely, for bundles with compact base space we have:

10.9 Proposition. *Let* $p : E \to X$ *and* $q : F \to X$ *be bundles of* Ω*-spaces with compact base space* X. *Then a mapping* $\lambda : E \to F$ *is an isomorphism if* Ω*-bundles if and only if*

(1) λ *is a homeomorphism.*

(2) λ *preserves stalks and* λ *is stalkwise a homomorphism of* Ω*-spaces.* □

10.10 Proposition. *Let* $p : E \to X$ *and* $q : F \to X$ *be bundles of* Ω*-spaces and assume that* $\Gamma(p)$ *and* $\Gamma(q)$ *are* Ω*-spaces. If the bundles* E *and* F *are isomorphic, then so are the* $C_b(X)$*-*Ω*-modules* $\Gamma(p)$ *and* $\Gamma(q)$. *The converse holds, provided that* $p : E \to X$ *and* $q : F \to X$ *are full bundles and provided that* X *is completely regular.*
In particular, this is the case if $\Gamma(p)$ *and* $\Gamma(p)$ *are complete metric spaces and if* X *is completely regular.*

Proof. It is only the converse which requires a proof.
Let us assume that $\Gamma(p)$ and $\Gamma(q)$ are complete metric spaces. Then all quotients of $\Gamma(p)$ and $\Gamma(q)$ are complete, too. From (2.6) we know that the evaluation maps $\varepsilon_x : \Gamma(p) \to p^{-1}(x)$ and $\varepsilon_x : \Gamma(q) \to q^{-1}(x)$ are quotient maps onto their images and hence the images are complete. As these images are also dense in the stalks, we conclude that the evaluation maps are surjections. Thus, the bundles $p : E \to X$ and $q : F \to X$ are full. Now apply (10 .7) to complete the proof. □

As a corollary we obtain the uniqueness of the bundle representing locally C(X)-convex C(X)-modules constructed in section 7:

10.11 Corollary. *Let* E *be a complete metrizable locally* $C_b(X)$*-con-*

convex $C(X)$-Ω-*module, where* X *is a compact space. Then, up to isomorphy, there is an unique bundle* $p : E \to X$ *of* Ω-*spaces such that* E *is isomorphic to* $\Gamma(p)$. □

Of course, all these results apply to bundles of Banach spaces. But dealing with Banach spaces, we always have to worry about the preservation of the norms, and this is what we shall do in the following remarks:

10.12 Definition. Let $p : E \to X$ and $q : F \to X$ be bundles of Banach spaces and let $\lambda : E \to F$ be a bundle morphism. We define

$$||\lambda|| = \sup \{ ||\lambda_{p^{-1}(x)}|| : x \in X\}. \quad \square$$

Note that by definition the maps $\lambda_{/p^{-1}(x)} : p^{-1}(x) \to q^{-1}(x)$ are bounded linear maps and that by (10.1(ii), property (d)) the number $||\lambda||$ is finite.

10.13 Proposition. *Let* $p : E \to X$ *and* $q : F \to X$ *be bundles of Banach spaces and let* $\lambda : E \to F$ *be a bundle morphism. Then*

$$||T_\lambda|| \leq ||\lambda||.$$

If all the evaluation maps $\varepsilon_x : \Gamma(p) \to p^{-1}(x)$ *are quotient maps of Banach spaces, then we have equality. This is especially the case if the base space* X *is completely regular.*

Proof. Let us compute: For all $\sigma \in \Gamma(p)$ we have

$$\begin{aligned} ||T_\lambda(\sigma)|| &= \sup \{ ||T_\lambda(\sigma)(x)|| : x \in X\} \\ &= \sup \{ ||\lambda(\sigma(x))|| : x \in X\} \\ &\leq \sup \{ ||\lambda_{/p^{-1}(x)}|| \cdot ||\sigma(x)|| : x \in X\} \end{aligned}$$

$$\leq \quad \|\lambda\| \cdot \|\sigma\| ,$$

whence $\|T_\lambda\| \leq \|\lambda\|$.

Conversely, assume that the evaluation map $\varepsilon_x : \Gamma(p) \to p^{-1}(x)$ is a quotient map of Banach spaces. Then for every bounded linear map $S : p^{-1}(x) \to F$ into an arbitrary Banach space F we have $\|S\| = \|S \circ \varepsilon_x\|$. Applying this to the equation $\varepsilon_x \circ T = \lambda_{/p^{-1}(x)} \circ \varepsilon_x$, we obtain the inequality

$$\begin{aligned} \|\lambda_{/p^{-1}(x)}\| &= \|\lambda_{/p^{-1}(x)} \circ \varepsilon_x\| \\ &= \|\varepsilon_x \circ T\| \\ &\leq \|\varepsilon_x\| \cdot \|T\| \\ &\leq \|T\| . \end{aligned}$$

This yields $\|\lambda\| \leq \|T\|$. □

10.14 Definition. Let $p : E \to X$ and $q : F \to X$ be bundles of Banach spaces and let $\lambda : E \to F$ be a morphism of bundles. If $\lambda_{/p^{-1}(x)} : p^{-1}(x) \to q^{-1}(x)$ is an isometry for each $x \in X$, then λ is called on *isometry of bundles*. If in addition λ is a bijection, then λ is called an *isometrical isomorphism of bundles*. □

In this definition we do not require an isometrical isomorphism to be open. But using axiom (1.5.II) it is very easy to show that this is always the case. Hence we have

10.15 Proposition. *Let* $p : E \to X$ *and* $q : F \to X$ *be bundles of Banach spaces. Then every isometrical isomorphism is an isomorphism in the sense of* (10.8). *If* $\lambda : E \to F$ *is an isometry (isometrical isomorphism) of bundles, then* $T_\lambda : \Gamma(p) \to \Gamma(q)$ *is an isometry (isometrical isomorphism) of Banach spaces.* □

For bundles with arbitrary base spaces, this is all I can say about norm preserving $C(X)$-module homomorphisms. To obtain better results, we have to consider bundles with completely regular base spaces:

10.16 Proposition. *Let* $p : E \to X$ *and* $q : F \to X$ *be bundles of Banach spaces with a completely regular base space* X *and let* $\lambda : E \to F$ *be a morphism of bundles. Then:*

(i) The operator T_λ *is an isometry if and only if* λ *is an isometry of bundles.*

(ii) The operator T_λ *is an isometrical isomorphism of Banach spaces if and only if* λ *is an isometrical isomorphism of bundles.*

Proof. (i): Suppose that $T_\lambda : \Gamma(p) \to \Gamma(q)$ is an isometry. We have to show that $||\lambda(\alpha)|| = ||\alpha||$ for every $\alpha \in E$.
Let $\alpha \in E$. From (10.13) we know that $||\lambda_{/p^{-1}(p(\alpha))}|| \leq ||\lambda|| = ||T_\lambda|| = 1$, whence $||\lambda(\alpha)|| \leq ||\alpha||$. To verify the converse inequality, we recall from (2.10) that the bundle $p : E \to X$ is full. Therefore we can find a section $\sigma \in \Gamma(p)$ such that $\sigma(p(\alpha)) = \alpha$. Now suppose that there is an $\varepsilon > 0$ such that $||\lambda(\alpha)|| + \varepsilon < ||\alpha||$. Then we also have $||T_\lambda(\sigma)(p(\alpha))|| = ||\lambda(\alpha)|| < ||\alpha|| - \varepsilon$. As norm : $F \to \mathbb{R}$ is upper semicontinuous, there is an open neighborhood U of $p(\alpha)$ such that $||T_\lambda(\sigma)(x)|| < ||\alpha|| - \varepsilon$ for all $x \in U$. As usual, we take a continuous function $f : X \to [0,1]$ such that $f(p(\alpha)) = 1$ and $f(X \setminus U) = \{0\}$. Then we conclude that $||f\cdot\sigma|| = ||T_\lambda(f\cdot\sigma)|| = ||f\cdot T_\lambda(\sigma)|| \leq ||\alpha|| - \varepsilon$, which is impossible as $||\alpha|| = ||f\cdot\sigma(p(\alpha))|| \leq ||f\cdot\sigma||$. Because $\varepsilon > 0$ was arbitrary, we have shown that $||\alpha|| \leq ||\lambda(\alpha)||$.
The other implication follows from (10.15)
(ii): One implication is again clear by (10.15). Thus, suppose

that T_λ is an isometrical isomorphism. Then λ is an isometry of bundles by (ii) and λ is a bijection, as a straightforward proof using (10.7) shows. Thus, λ is an isometrical isomorphism. □

We collect all these partial results:

10.18 Summary. *Let* $p : E \to X$ *and* $q : F \to X$ *be bundles of Banach spaces with a completely regular base space. Then the mapping*

$$\lambda \to T_\lambda, \qquad T_\lambda(\sigma) = \lambda \circ \sigma$$

is a bijection between the set of all bundle morphisms $\lambda : E \to F$ *and the set of all bounded* $C_b(X)$*-module homomorphisms* $T : \Gamma(p) \to \Gamma(q)$. *The inverse of the mapping is given by*

$$T \to \lambda_T$$

where $\lambda_{T/p^{-1}(x)} : p^{-1}(x) \to q^{-1}(x)$ *is the unique map such that the diagram*

$$\begin{array}{ccc} \Gamma(p) & \xrightarrow{T} & \Gamma(q) \\ \varepsilon_x \downarrow & & \downarrow \varepsilon_x \\ p^{-1}(x) & \xrightarrow[\lambda_T]{} & q^{-1}(x) \end{array}$$

commutes.

Moreover, the mapping $\lambda \to T_\lambda$ *preserves norms, sends isometries of bundles onto isometries of Banach spaces and isometrical isomorphisms of bundles onto isometrical isomorphisms of Banach spaces.* □

10.19 Remarks. a) If $\lambda : E \to F$ is injective, then so is

$T_\lambda : \Gamma(p) \to \Gamma(q)$. Example (10.20) shows that the converse is false.

b) From (10.7) and Banach's homomorphism theorem we may conclude that for bundles of Banach spaces with a completely regular base space, a bundle morphism $\lambda : E \to F$ is bijective whenever the operator $T_\lambda : \Gamma(p) \to \Gamma(q)$ is bijective. Example (10.24) shows that the converse does not even hold for bundles with compact base spaces.

c) Let us again consider bundles with completely regular base spaces and let $\lambda : E \to F$ be a bundle morphism. If $T_\lambda : \Gamma(p) \to \Gamma(q)$ is a surjection, then we may see that λ is onto, too. Conversely, if λ is onto, then the image of T_λ has not even to be dense in $\Gamma(q)$ (see example (10.25). However, using the Stone-Weierstraß theorem (4.3), $T_\lambda(\Gamma(p))$ is dense in $\Gamma(q)$ whenever the base space X is compact and λ is surjective. Again, example (10.24) will show that this is all we can expect.

10.20 Example. Let X be an arbitrary topological space and let E and F be topological vector spaces. Then we consider the trivial bundles $p : X\times E \to X$ and $q : X\times F \to X$, where p and q are the first projections. We know from (1.8) that $\Gamma(p) = C_b(X,E)$ and $\Gamma(q) = C_b(X,F)$, where $C_b(X,E)$ (resp. $C_b(X,F)$) denotes the topological vector space of all E-valued (F-valued) bounded continuous functions, equipped with the topology of uniform convergence on X.
We shall give a description of all bundle morphisms $\lambda : X\times E \to X\times F$. If X is completely regular, this will yield a description of all $C_b(X)$-module homomorphisms from $C_b(X,E) \to C_b(X,F)$.

Let us start with a bundle morphism $\lambda : X\times E \to X\times F$. Then the restriction of λ to $\{x\}\times E$ is linear and continuous. Hence for every $x \in X$

there is a continuous linear mapping $\lambda_x : E \to F$ such that

$$(x,a) = (x,\lambda_x(a)) \qquad \text{for every } (x,a) \in X\times E$$

Thus we have a mapping $\lambda_- : X \to L_s(E,F)$, where $L_s(E,F)$ denotes the space of all continuous linear mappings from E into F, equipped with the topology of pointwise convergence.

The mapping $\lambda_- : X \to L_s(E,F)$ is continuous: Indeed, for every $a \in E$ the mapping $x \to \lambda_x(a) : X \to F$ is continuous, as this mapping is the composition of $x \to (x,a) \to \lambda(x,a) = (x,\lambda_x(a)) \to \lambda_x(a)$. As the topology on $L_s(E,F)$ is the topology of pointwise convergence, the continuity of λ_- follows.

Moreover, the set $\{\lambda_x : x \in X\}$ is equicontinuous: Take any continuous seminorm ω on F. We have to find an open neighborhood $U \subset E$ of O such that $\omega(\lambda_x(a)) \leq 1$ for all $x \in X$ and all $a \in U$.
Firstly, define a seminorm $\mu : X\times F \to \mathbb{R}$ on the bundle $q : X\times F \to X$ by setting $\mu(x,b) = \omega(b)$. By (1.8(i)) we may think of μ as one of the seminorms belonging to the bundle $q : X\times F \to X$. As $\lambda : X\times E \to X\times F$ is a morphism of bundles, there is a real number $M > 0$ and a seminorm $\nu : X\times E \to \mathbb{R}$ of the bundle $p : X\times E \to X$ such that $\nu(x,a) \leq M$ implies $\omega(\lambda_x(a)) = \mu(x,\lambda_x(a)) = \mu(\lambda(x,a)) \leq 1$. Again by (1.8(i)), we can find a continuous seminorm $\kappa : E \to \mathbb{R}$ such that $\nu(x,a) = \kappa(a)$ for all $a \in E$. Now let $U = \{a \in E : \kappa(a) < M\}$. Then U is an open neighborhood of $0 \in E$. Furthermore, $a \in U$ implies $\omega(\lambda_x(a)) \leq 1$ for all $x \in X$. This shows the equicontinuity of the set $\{\lambda_x : x \in X\}$.

Conversely, let $\lambda_- : X \to L_s(E,F)$ be a continuous function such that the image $\{\lambda_x : x \in X\}$ is equicontinuous. We define a mapping

$$\lambda : X\times E \to X\times F \qquad \text{by}$$
$$(x,a) \to (x,\lambda_x(a)) .$$

Then λ is continuous: It is enough to show that $(x,a) \to \lambda_x(a)$: $X\times E \to F$ is continuous: Let $(x_o,a_o) \in X\times E$ and let W be a neighborhood of $\lambda_{x_o}(a_o)$. Pick any neighborhood V of $0 \in F$ such that $\lambda_{x_o}(a_o) + V \subset W$ and let U be any neighborhood of $0 \in F$ such that $U + U \subset V$. As the mapping $x \to \lambda_x(a_o)$ is continuous, we may find an open neighborhood S of x_o such that $\lambda_x(a_o) \in \lambda_{x_o}(a_o) + U$ for all $x \in S$. Moreover, the equicontinuity of $\{\lambda_x : x \in X\}$ yields an open neighborhood $T \subset E$ of 0 such that $\lambda_x(T) \subset U$ for all $x \in X$. Thus, for $(x,a_o + t) \in S\times(a_o + T)$ we have $\lambda_x(a_o + t) = \lambda_x(a_o) + \lambda_x(t) \in$
$\subset \lambda_{x_o}(a_o) + U + U \subset \lambda_{x_o}(a_o) + V \subset W$.
It is now obvious that λ satisfies the properties (a), (b) and (c) of definition (10.1(ii)). We check property (d):
Let $\mu : X\times F \to F$ be one of the seminorms of the bundle $q : X\times F \to X$. By (1.8(i)) we may assume that $\mu(x,b) = \omega(b)$, where ω is a certain continuous seminorm on F. Again, we make use of the equicontinuity of the set $\{\lambda_x : x \in X\}$ to find a continuous seminorm κ on E and a number $M > 0$ such that $\kappa(a) \leq M$ implies $\omega(\lambda_x(a)) \leq 1$ for all $x \in X$. Now the mapping $\nu : X\times E \to \mathbb{R}$ defined by $\nu((x,a)) = \kappa(a)$ is a continuous seminorm on $X\times E$ and $\nu((x,a)) \leq M$ implies $\mu(\lambda(x,a)) \leq 1$.

Thus, λ is a bundle morphism, and we have shown

10.21 Let X be a topological space and let E,F be topological vector spaces. Then the mapping

$$(\lambda : X\times E \to X\times F) \quad \to \quad (\lambda_- : X \to L_s(E,F))$$

is a bijection between the set of all bundle morphisms from $X\times E$ into $X\times F$ and the set of all continuous mappings from X into $L_s(E,F)$ such that the image is equicontinuous.

Of course, if we wish to consider an additional Ω-structure on E and F, then we have to replace $L_s(E,F)$ by the subspace of all Ω-homomorphisms.

In certain cases every bounded subset of $L_s(E,F)$ is already equicontinuous. This is for instance so, if E is a Baire space and especially if E is a Banach space (see [Sch 71, theorem III.4.2]). Thus, we can state:

10.22 Let X be a topological space and let E and F be topological vector spaces such that E is a Baire space. Then the mapping

$$(\lambda : X\times E \to X\times F) \quad \to \quad (\lambda_- : X \to L_s(E,F))$$

is a bijection between the set of all bundle morphisms and $C_b(X,L_s(E,F))$.

We may interpret (10.22) as a bundle representation of the set of all bundle morphisms between $X\times E$ and $X\times F$. We shall return to this idea in a later section, when we discuss bundles of operators.

Combining (10.7) and (10.21) we obtain:

10.23 Let X be a completely regular space and let E and F be topological vector spaces (such that E is a Baire space). Then a continuous operator $T : C_b(X,E) \to C_b(X,F)$ is a $C_b(X)$-module homomorphism if and only if there is a continuous mapping $\lambda : X \to L_s(E,F)$

such that $\lambda(X)$ is equicontinuous (bounded) and $T(\sigma)(x) = \lambda(x)(\sigma(x))$ for all $x \in X$ and all $\sigma \in C_b(X,E)$.

Concretely, we take $X = [0,1]$ with its usual topology and $E = F = \mathbb{R}$. Then the $C(X)$-module homomorphisms from $C([0,1])$ into $C([0,1])$ are given by multiplication with continuous functions $f \in C([0,1])$ (as everybody knows). If we take the mapping $id : x \to x \in C([0,1])$, then the $C(X)$-module homomorphism

$$T : C([0,1]) \to C([0,1])$$
$$f \to id \cdot f$$

is injective, but the corresponding bundle homomorphis $\lambda_T : X \times \mathbb{R} \to$ $\to X \times \mathbb{R}$ is given by

$$\lambda_T : X \times \mathbb{R} \to X \times \mathbb{R}$$
$$(x,r) \to (x,r \cdot x)$$

and thus is not injective on the fiber over $x = 0$.

10.24 Example. Let $p : [0,1] \times \mathbb{R} \to [0,1]$ be the bundle constructed in (5.16). Recall that $[0,1] \times \mathbb{R}$ does not carry the product topology and that $\Gamma(p)$ is the completion of $C([0,1])$ in the norm

$$|||f||| = \max \{|f(0)|, \sup \{x \cdot |f(x)| : 0 < x \leq 1\}\}.$$

Recall also that the canonical injection $T : C([0,1]) \to \Gamma(p)$ is a $C([0,1])$-module homomorphism which is not surjective. Nevertheless, the corresponding bundle morphism $\lambda_T : [0,1] \times \mathbb{R} \to [0,1] \times \mathbb{R}$ from the trivial bundle $pr_1 : [0,1] \times \mathbb{R} \to [0,1]$ into $p : [0,1] \times \mathbb{R} \to [0,1]$ is the identity map and therefore a bijection.

10.25 Example. This time we take as base space X the whole real line and consider the trivial bundle $pr_1 : \mathbb{R} \times \mathbb{R} \to \mathbb{R}$. In this case

$\Gamma(pr_1) = C_b(\mathbb{R})$. As an operator $T : C_b(\mathbb{R}) \to C_b(\mathbb{R})$ we take multiplication with the continuous function $\exp(-x^2)$. Then T maps $C_b(\mathbb{R})$ into the closed subspace $C_o(\mathbb{R})$ of all continuous functions on $\mathbb{R}$ vanishing at infinity. Thus, T is not surjective.
In this case again, the corresponding bundle map $\lambda_T : \mathbb{R} \times \mathbb{R} \to \mathbb{R} \times \mathbb{R}$ is even a homeomorphism.

11. Bundles of operators

In this section we shall study spaces of continuous operators into the space of sections of a bundle. The basic ideas may be explained with the following example:

Let E be a normed space and let X be a compact space. By $K(E,C(X))$ we denote the Banach space of all compact operators from E into $C(X)$. It is well-known that $K(E,C(X))$ is isometrically isomorphic with the Banach space $C(X,E')$ of all norm-continuous mappings from X into E', equipped with the supremum norm. The canonical isomorphism

$$\Phi \ : \ K(E,C(X)) \ \to \ C(E,E')$$

is given by

$$\Phi(u)(x) = \varepsilon_x \circ u$$

where $\varepsilon_x : C(X) \to \mathbb{K}$ is the usual evaluation map.

Hence, we have obtained a bundle representation of the space of all compact operators. The stalks of this bundle are all identical with $E' = L_b(E, \mathbb{K})$, i.e. they may be viewed as the set of all bounded operators from E into the stalks of the trivial bundle $pr_1 : X \times \mathbb{K} \to X$, equipped with the topology of uniform convergence on bounded sets.

Unfortunately, this example also shows that we can not expect such a nice representation in general: Let us try to present the Banach space $L_b(E,C(X))$ of all bounded operators with the operator norm as a space of sections in a bundle of Banach spaces. If X is infinite, then $L_b(E,C(X))$ is strictly larger than $K(E,C(X))$; hence $L_b(E,C(X))$

cannot be represented in the form $C(X,E')$, where E' carries the norm topology. However, it is known that $L_s(E,C(X))$ equipped with the topology of pointwise convergence is topological isomorphic to $C(X,E'_s)$, where E'_s carries the $\sigma(E',E)$-topology. This shows that we have to choose an appropriate topology in order to obtain a "nice" representation of $L(E,C(X))$ by sections in a bundle.

On the other hand, $L_b(E,C(X))$, equipped with the operator norm, may indeed be written as the space of all sections in a bundle of Banach spaces with base space X. However, the stalks of this bundle are not as nice as they are in the other bundle. We shall give a rather technical description of them.

To start our discussion, we recall some facts concerning topologies on the space $L(E,F)$ of all continuous linear operators from E into F (see [Sch 71, III.3]):

Let E and F be topological vector spaces and let $\mathcal{S}$ be a family of bounded subsets of E such that the linear hull of $\cup\mathcal{S}$ is dense in E (a family with the second property is called *total*). If we equip $L(E,F)$ with the topology of uniform convergence on all subset $S \in \mathcal{S}$ of E, then $L(E,F)$ becomes a locally convex Hausdorff topological vector space. A base of open neighborhoods of 0 for this topology is given by sets of the form

$$\mathcal{U}(S,U) := \{T \in L(E,F) : T(S) \subset U\}$$

where S runs through all elements of $\mathcal{S}$ and U ranges over an open neighborhood base of $0 \in F$.

If the topology of F is generated by a family of seminorms $(\nu_j)_{j\in J}$, then the topology of uniform convergence on subsets in $\mathcal{S}$ is generated by the family of seminorms $(\nu_{S,j})_{S\in\mathcal{S},j\in J}$ given by

$$\nu_{S,j}(T) = \sup_{u\in S} \nu_j(T(u))$$

If the family $(\nu_j)_{j\in J}$ is directed and we want to have the same property for the family $(\nu_{S,j})_{S\in\mathcal{S},j\in J}$, we have to require that the family $\mathcal{S}$ is directed in the sense that for every pair $S_1, S_2 \in \mathcal{S}$ there is an element $S_3 \in \mathcal{S}$ such that $S_1 \cup S_2 \subset S_3$. This is the case for the following examples:

a) The topology of pointwise convergence: $\mathcal{S}$ is the family of all finite subsets of E and we denote by $\mathcal{L}_s(E,F)$ the space $\mathcal{L}(E,F)$ equipped with this topology.

b) The topology of compact convergence: $\mathcal{S}$ is the family of all compact subsets of E; the corresponding space is denoted by $\mathcal{L}_c(E,F)$.

c) The topology of compact, convex circled convergence, provided that E is quasicomplete: $\mathcal{S}$ is the family of all compact, convex circled subsets of E; the space of operators with this topology is denoted by $\mathcal{L}_{cc}(E,F)$.

d) The topology of precompact convergence: $\mathcal{S}$ consists of all precompact subsets; the space is denoted by $\mathcal{L}_{pc}(E,F)$.

e) The topology of bounded convergence: $\mathcal{S}$ consists of all bounded subsets; the space of operators is denoted by $\mathcal{L}_b(E,F)$.

f) In general, we denote the space $\mathcal{L}(E,F)$ equipped with the topology of uniform convergence on all subsets $S \in \mathcal{S}$ by $\mathcal{L}_{\mathcal{S}}(E,F)$.

Note that $\mathcal{L}_b(E,F)$ is a normed space provided that E and F are normed spaces. In this case we may take $\mathcal{S} = \{B_1(E)\}$. Moreover, the corresponding seminorm is the operator norm.

Now let us suppose that F is a topological $C_b(X)$-module. Then we may define a multiplication with elements of $C_b(X)$ on $L(E,F)$ in the following way:

$$(f\circ T)(u) := f\circ(T(u)).$$

It is obvious that $L(E,F)$ will be a $C_b(X)$-module under this multiplication. Moreover, we have:

11.1 Proposition. *Let* E *be a topological vector space and let* F *be a topological* $C_b(X)$*-module.*

(i) *If* $S \subset E$ *is any subset and if* $U \subset F$ *is a* $C_b(X)$*-convex subset, then* $\{T \in L(E,F) : T(S) \subset U\}$ *is* $C_b(X)$ *-convex.*

(ii) *If* S *is any family of bounded subsets of* E *and if* F *is a locally* $C_b(X)$*-convex* $C_b(X)$*-module, then* $L_S(E,F)$ *is a locally* $C_b(X)$*-convex* $C_b(X)$*-module, too.* □

If E and F are Banach spaces and if F is a locally $C_b(X)$-convex $C_b(X)$-module as a normed space (recall that for normed spaces the $C_b(X)$-convexity means that the closed unit ball is $C_b(X)$-convex), then we may apply (11.1(i)) with $S = B_1(E)$ and $U = B_1(F)$ to obtain:

11.2 Corollary. *Let* E *and* F *be Banach spaces and let us assume that* E *is a locally* $C_b(X)$*-convex* $C_b(X)$*-module. Then* $L_b(E,F)$*, equipped with the operator norm, is a locally* $C_b(X)$*-convex* $C_b(X)$*-module, too.* □

Hence, applying (7.16) we learn the following:

11.3 Corollary. *If* $p : E \to X$ *is a bundle with a quasicompact base spase and if* F *is a topogical vector space, then there is a bundle*

$q : F \to X$ *such that* $L_S(F,\Gamma(p))$ *is isomorphic to a* $C(X)$*-submodule of* $\Gamma(q)$, *provided that* S *is total in* F. □

Our first problem will be to identify the stalks of the bundle $q : F \to X$. If we look at the examples at the beginning, we would hope that they are at least subspaces of $L_S(F,p^{-1}(x))$. If we recall the construction of the stalks (see section 7), it seems to be reasonable to restrict ourselves to completely regular base spaces X, as otherwise in might happen that every $\mathbb{K}$-valued continuous function is constant.In this case the construction in section 7 leads to bundles whose stalks are isomorphic to the whole space, which certainly is no progress at all.

The second problem then will be to decide whether or not $L_S(F,\Gamma(p))$ is not only dense in $\Gamma(q)$ but even equal to $\Gamma(q)$. For compact base spaces, a first answer is

11.4 Proposition. *Let* $p : E \to X$ *be a bundle with a compact base space* X *such that all stalks are complete. If* F *is a bornological space and a the family* S *of subsets of* F *contains the closure of every* 0*-sequence, then there is a bundle* $q : F \to X$ *such that* $L_S(F,\Gamma(p))$ *is isomorphic to* $\Gamma(q)$.

Proof. We know from (1.10) that $\Gamma(p)$ is complete. Hence, we may deduce from [Sch 71, p.117, exercise 8] that $L_S(F,\Gamma(p))$ is complete. The proposition is now an easy consequence of (7.16). □

Before we get to work and identify the stalks of the bundle $q : F \to X$, we close our general discussion with a corollary:

11.5 Corollary. *Let* $p : E \to X$ *be a bundle of Banach spaces and let*

F *be a Banach space. Then the spaces* $L_c(F,\Gamma(p))$ *and* $L_b(F,\Gamma(p))$ *may both be represented as the space of all sections in a bundle* $q_c : F_c \to X$ *and* $q_b : F_b \to X$ *resp.* □

From now on we shall pass to a slightly more general situation: We shall always consider a bundle $p : E \to X$ such that the base space X is at least completely regular. Moreover, L will always denote a $C_b(X)$-submodule of $L(F,\Gamma(p))$. Finally, S will be a family of bounded subsets such that $F = \langle \cup S \rangle$ and such that $T(S)$ is precompact for every $T \in L$ and every $S \in S$. The space L will always carry the topology of uniform convergence on subsets $S \in S$, i.e. the relative topology inherited from $L_S(F,\Gamma(p))$.

11.6 Proposition. *Under the above assumptions, the closure in* L *of the set* $I_x \cdot L = \{f \cdot T : f \in C_b(X),\ f(x) = 0,\ T \in L\}$ *is equal to* $\{T \in L : T(F) \subset N_x\}$, *where* $N_x = \{\sigma \in \Gamma(p) : \sigma(x) = 0\}$.

Proof. Let $T \in L$ and let $f \in C_b(X)$ such that $f(x) = 0$. Then for every $a \in F$ we have $(f \cdot T)(a)(x) = f(x) \cdot (T(a)(x)) = 0$, i.e. $f \cdot T(a) \in N_x$. This implies $T(F) \subset N_x$.
Moreover, the set $\{T : T(F) \subset N_x\}$ is closed: Indeed, let $(T_i)_{i \in I}$ be any convergent net contained in $\{T \in L : T(F) \subset N_x\}$. Then this net is also convergent in the topology of pointwise convergence. But the set $\{T \in L : T(F) \subset N_x\}$ is obviously closed in the topology of pointwise convergence. Hence $\{T \in L : T(F) \subset N_x\}$ is closed in L.

It remains to show that $I_x \cdot L$ is dense in $\{T \in L : T(F) \subset N_x\}$. To prove this, let $T \in L$ be any continuous operator such that $T(F) \subset N_x$, let $S \in S$ be any element, let $\varepsilon > 0$ and let ν_j be one of the seminorms belonging to the bundle $p : E \to X$. It suffices to find an function

$g \in C_b(X)$ such that $g(x) = 0$ and $\sup_{s \in S} \hat{\nu}_j((1 - g) \cdot T(s)) \leq \varepsilon$.

As $T(S)$ is precompact, we can find $a_1, \ldots, a_n \in S$ such that for every $a \in S$ there is an $i \in \{1, \ldots, n\}$ such that $\hat{\nu}_j(T(a_i - a)) < \varepsilon/2$. As $T(a_i)(x) = 0$ for all $1 \leq i \leq n$ and as $\nu_j : E \to \mathbb{R}$ is upper semicontinuous, we can find an open neighborhood U of x such that $\nu_j(T(a_i)(y)) < \varepsilon/2$ for all $y \in U$ and all $1 \leq i \leq n$. Use the fact that X is completely regular to find a continuous mapping $g : X \to [0,1]$ satisfying $g(x) = 0$ and $g(X \setminus U) = \{1\}$. By standard arguments we obtain $\hat{\nu}_j((1 - g)(T(a_i)) \leq \varepsilon/2$ for all $1 \leq i \leq n$. If $a \in S$ is arbitrary, then there is a certain $i \in \{1, \ldots, n\}$ such that $\hat{\nu}_j((1-g) \cdot T(a - a_i)) \leq$ $\leq \hat{\nu}_j(T(a - a_i)) \leq \varepsilon/2$. Now the triangle inequality yields $\hat{\nu}_j((1 - g) \cdot T(a)) \leq \varepsilon$ for all $a \in S$. Hence we have

$$\sup_{s \in S} \hat{\nu}_j((1 - g) \cdot T(s)) \leq \varepsilon,$$

as desired. □

In the following, we shall again make use of our convention to denote the stalks of the bundle $p : E \to X$ by E_x, $x \in X$. Further, we let $N_x = \{\sigma \in \Gamma(p) : \sigma(x) = 0\}$ and $\varepsilon_x : \Gamma(p) \to E_x$ be the evaluation map.

If L is a subspace of $L_S(F, \Gamma(p))$, we define

$$N_x^L = \{T \in L : T(F) \subset N_x\}$$

and

$$L_x = L/N_x^L \qquad \text{equipped with the quotient topology.}$$

In the following, we shall give a description of a family of seminorms generating the topology on L_x:

Let ν_j be on of the seminorms of the bundle $p : E \to X$ and let $S \in \mathcal{S}$. We define:

$$\nu_{S,j}^{x}(T + N_x^L) := \inf \{\sup_{a\in S} \nu_j(T(a) + T'(a)) : T' \in N_x^L\}$$

Then the seminorms $(\nu_{S,j}^{x})_{S,j}$ will generate the topology on L_x. Nobody can work with such a formula, therefore we give an alternative expression for these seminorms:

11.7 Proposition. *Under the assumptions made in the remarks preceeding (11.6), we have*

(i) *The mapping* $T + N_x^L \to \varepsilon_x \circ T : L_x \to L(F,E_x)$ *is well defined, linear and injective.*

(ii) *For every seminorm* $\nu_j : E \to \mathbb{R}$ *of the bundle* $p : E \to X$, *every* $S \in S$, *every* $x \in X$ *and every* $T \in L$ *we have*

$$\nu_{S,j}^{x}(T + N_x^L) = \sup_{a\in S} \nu_j(\varepsilon_x \circ T(a))$$

(iii) *In particular, the mapping* $T + N_x^L \to \varepsilon_x \circ T : L_x \to L(F,E_x)$ *is an embedding.*

Proof. The property (i) is a consequence of (11.6) and (iii) follows immediately from (ii). Thus, it remains to check (ii):

First of all, for every $T' \in N_x^L$ we have

$$\begin{aligned}\sup_{a\in S} \hat{\nu}_j(T(a) + T'(a)) &= \sup_{a\in S}\ \sup_{y\in X} \nu_j(T(a)(y) + T'(a)(y)) \\ &\geq \sup_{a\in S} \nu_j(T(a)(x) + T'(a)(x)) \\ &= \sup_{a\in S} \nu_j(T(a)(x)) \qquad \text{(since } T'(a)(x) = 0\text{)} \\ &= \sup_{a\in S} \nu_j(\varepsilon_x \circ T(a)) ,\end{aligned}$$

and therefore $\nu_{S,j}^{x}(T + N_x^L) \geq \sup_{a\in S} \nu_j(\varepsilon_x \circ T(a))$ by the definition of the $(\nu_{S,j}^{x})_{S,j}$.

Conversely, suppose that there is a $C > 0$ such that

$$\vartheta^x_{S,j}(T + N^L_x) \quad > \quad C \quad > \quad \sup_{a \in S} \nu_j(\varepsilon_x \circ T(a)).$$

In this case, we let

$$\varepsilon := \frac{1}{2}(C - \sup_{a \in S} \nu_j(\varepsilon_x \circ T(a))) .$$

As the set $T(S) \subset \Gamma(p)$ is precompact, we may find elements $a_1, \dots, a_n \in S$ such that for every $a \in S$ there is an index $i \in \{1, \dots, n\}$ with $\vartheta_j(T(a) - T(a_i)) < \varepsilon$ and we conclude that $\nu_j(\varepsilon_x \circ T(a_i)) < C - \varepsilon$ for all $1 \le i \le n$.

Now the upper semicontinuity of the mappings $y \to \nu_j(T(a_i)(y)) : X \to \mathbb{R}$ yields an open neighborhood U of x such that $\nu_j(T(a_i)(y)) < C - \varepsilon$ for all $y \in U$ and all $i \in \{1, \dots, n\}$. Choose a continuous function $g : X \to [0,1]$ such that $g(x) = 0$ and $g(X \setminus U) = \{1\}$. Then we have

$$\begin{aligned} \vartheta_j((1 - g) \cdot T(a_i)) \quad &= \quad \sup_{y \in X} \nu_j((1 - g(y)) \cdot T(a_i)(y)) \\ &\le \quad C - \varepsilon \end{aligned}$$

for all $1 \le i \le n$.

If $a \in S$ is arbitrary, then $\vartheta_j(T(a_i) - T(a)) < \varepsilon$ for a certain $i \in \{1, \dots, n\}$. Hence the triangle inequality yields

$$\vartheta_j((1 - g) \cdot T(a)) \quad \le \quad C$$

and therefore

$$\sup_{a \in S} \vartheta_j(T(a) - g \cdot T(a)) \le C.$$

From (11.6) we conclude that $-g \cdot T \in N^L_x$. This leads to the contradiction

$$\begin{aligned} C \quad &< \quad \vartheta^x_{S,j}(T + N^L_x) \\ &\le \quad \sup_{a \in S} \nu_j(T(a) - g \cdot T(a)) \\ &\le \quad C. \quad \square \end{aligned}$$

11.8 Proposition. *Under the same assumptions, we have*

(i) *The mapping* $x \to \sup_{a\in S} \nu_j(\varepsilon_x \circ T(a)) : X \to \mathbb{R}$ *is upper semicontinuous for every* $T \in L$, $S \in \mathcal{S}$ *and every seminorm* $\nu_j : E \to \mathbb{R}$ *of the bundle* $p : E \to X$.

(ii) $\sup_{a\in S} \hat{\nu}_j(T(a)) = \sup_{x\in X} \sup_{a\in S} \nu_j(\varepsilon_x \circ T(a))$.

Proof. Using (7.7), we conclude that the mapping $x \to \hat{\nu}^x_{S,j}(T + N^L_x)$ is upper semicontinuous. Thus, (i) follows from (11.7).
The proof of (ii) is an easy calculation:

$$\begin{aligned}\sup_{a\in S} \hat{\nu}_j(T(a)) &= \sup_{a\in S} \sup_{x\in X} \nu_j(T(a)(x)) \\ &= \sup_{x\in X} \sup_{a\in S} \nu_j(\varepsilon_x \circ T(a)). \qquad \square\end{aligned}$$

We are now in the position to prove a bundle representation of $C(X)$-submodules $L \subset L_S(F,\Gamma(p))$. Our first result is still rather technical:

11.9 Proposition. *Let* $p : E \to X$ *be a bundle with a completely regular base space and seminorms* $(\nu_j)_{j\in J}$ *and let* F *be a topological vector space. Further, let* $L \subset L_S(F,\Gamma(p))$ *be a* $C_b(X)$*-submodule, where* S *is a directed family of bounded subsets of* F *such that* $F = \langle \cup S \rangle$ *and such that* $T(S)$ *is precompact in* $\Gamma(p)$ *for every* $S \in \mathcal{S}$ *and every* $T \in L$.
Then there is a full bundle $q_L : F_L \to X$ *such that* L *is isomorphic to a* $C_b(X)$*-submodule of* $\Gamma(q_L)$. *The stalk over* $x \in X$ *of this bundle may be chosen to be a subspace of* $L_S(F,E_x)$, *where* E_x *is the stalk over* x *of the bundle* $p : E \to X$. *In this case, the canonical injection*

$$\Phi : L \to \Gamma(q_L)$$

is given by $\Phi(T)(x) = \varepsilon_x \circ T$, *where* $\varepsilon_x : \Gamma(p) \to E_x$ *is the canonical evaluation.*

Proof. For every $x \in X$ let $M_x = \{\varepsilon_x \circ T : T \in L\} \subset L_S(F,E_x)$. If $(\nu_j)_{j\in J}$ is the family of seminorms of the bundle $p : E \to X$, then the topology on M_x is induced by the seminorms $(\omega^x_{S,j})_{(S,j)\in S\times J}$ given by

$$\omega^x_{S,j}(\alpha) = \sup_{u\in S} \nu_j(\alpha(u)) \qquad ,\alpha \in M_x.$$

Moreover, by (11.8(ii)), the space L may be identified with a subspace of $\prod^\infty_{x\in X} M_x$. The embedding $L \to \prod^\infty_{x\in X} M_x$ is given by

$$T \to \Phi(T)$$

$$\Phi(T)(x) = \varepsilon_x \circ T$$

It is now easy to verify that L, viewed as a subspace of $\prod^\infty_{x\in X} M_x$, satisfies the axioms (FM3) and (FM4) of section 5. Therefore an application of (5.8) completes the proof. □

In general, there is no reason to believe that L is isomorphic to the space of <u>all</u> section of $\Gamma(q_L)$. For example, let X be compact and let $N \subset L_S(F,C(X))$ be the space of all nuclear operators from a normed space F into C(X), equipped with the operator norm. As every nuclear operator is compact, the above result applies to N and we obtain a bundle $q_N : F_N \to X$ such that N may be identified with a C(X)-submodule of $\Gamma(q_N)$. The Stone-Weierstraß theorem implies that N is dense in $\Gamma(q_N)$ and it turns out that $\Gamma(q_N)$ is isomorphic to the space of all compact operators, i.e. N is strictly contained in $\Gamma(q_N)$.

Thus, it is of some interest to study the space of all sections of $\Gamma(q_L)$. It turns out that every section of the bundle $q_L : F_L \to X$ may be viewed as a linear operator from F into $\Gamma(p)$, but these operators will not be continuous in general.

11.10 Proposition. *Let* $p : E \to X$, $L \subset L_S(F,\Gamma(p))$ *and* S *be as in* (11.9).

(i) *If* $\Sigma \in \Gamma(q_L)$ *is a continuous section, then*

$$T_\Sigma : F \to \Gamma(p)$$

defined by

$$T_\Sigma(u)(x) = \Sigma(x)(u) \qquad \text{for all } x \in X \text{ , all } u \in F$$

is a linear map between F *and* $\Gamma(p)$.

(ii) *Under each of the following conditions, the mapping* T_Σ *is continuous:*

(a) S contains a neighborhood of $0 \in F$.

(b) F is bornological, X is compact and S contains the closure of every 0-sequence in F.

In these cases, the mapping $\Sigma \to T_\Sigma : \Gamma(q_L) \to L_S(F,\Gamma(p))$ *is an embedding.*

11.11 Remarks (i) If we compare the case (b) of this proposition with (11.4), we see that we may drop the completeness of the stalks in the hypothesis of (11.4).

(ii) We shall see in the following proof that T_Σ will be always sequentially continuous, provided that X is compact and that S contains the closure of every 0-sequence in F.

Proof of (11.10). (i) : Obviously, the mapping T_Σ will be linear. Whence it is enough to show that T_Σ maps F into $\Gamma(p)$.
Thus, let us start with $u_o \in F$. As the family of sets S generates F, we may assume that $u_o \in S_o$ for a certain $S_o \in S$.
Firstly, we show that $T_\Sigma(u_o)$ is bounded: Let ν_j be any of the seminorms of the bundle $p : E \to X$. If the seminorms $(\omega^x_{S,j})_{S,j}$ on the

bundle $q_L : F_L \to X$ are defined as in the proof of (11.9), then we may estimate:

$$
\begin{aligned}
\mathcal{V}_j(T_\Sigma(u_o)) &= \sup_{x\in X} \nu_j(T_\Sigma(u_o)(x)) \\
&= \sup_{x\in X} \nu_j(\Sigma(x)(u_o)) \\
&\le \sup_{u\in S_o} \sup_{x\in X} \nu_j(\Sigma(x)(u)) \\
&= \sup_{x\in X} \omega^x_{S_o,j}(\Sigma(x)) \\
&< \infty
\end{aligned}
$$

as Σ belongs to $\Gamma(q_L)$ and therefore is a bounded selection.

To show the continuity of the mapping $T_\Sigma(u_o) : X \to E$, we state the following

(*) Let $x_o \in X$ and let $T \in L$ be such that $\varepsilon_{x_o} \circ T = \Sigma(x_o)$. Then for every seminorm ν_j belonging to the bundle $p : E \to X$, every $\varepsilon > 0$ and every $S \in S$ there is an open neighborhood W of x_o such that $\nu_j(\Sigma(x)(u) - T(u)(x)) < \varepsilon$ for all $u \in S$ and all $x \in W$.

Indeed, the property (*) follows immediatlely from the upper semicontinuity of the mapping $x \to \omega^x_{S,j}(\Sigma(x) - \varepsilon_x \circ T) = \sup_{u\in S} \nu_j(\Sigma(x)(u) - T(u)(x))$.

Now (*) implies the continuity of $T_\Sigma(u_o)$ at x_o: Firstly, by the definition of the stalks of F_L (see the proof of (11.9)), we can pick an operator $T \in L$ such that $\varepsilon_{x_o} \circ T = \Sigma(x_o)$. In this case we have $T_\Sigma(u_o)(x_o) = T(u_o)(x_o)$ and as $T(u_o)$ belongs to $\Gamma(p)$, a typical open neighborhood V of $T_\Sigma(u_o)(x_o)$ looks like

$$V = \{\alpha \in E : p(\alpha) \in W', \nu_j(\alpha - T(u_o)(p(\alpha))) < \varepsilon\},$$

where W' is an open set around x_o. Now use (*) to find an open neighborhood W of x_o such that $\nu_j(\Sigma(x)(u_o) - T(u_o)(x)) < \varepsilon$ for all $x \in W$. Then by definition the mapping T_Σ maps the neighborhood $W \cap W'$ of x_o into V.

(ii): Case a. Let $U \in S$ be a neighborhood of $0 \in F$. As Σ belongs to $\Gamma(q_L)$, it is a bounded selection. Hence for every $j \in J$ the number

$$\begin{aligned}\sup_{u\in U} \hat{\nu}_j(T_\Sigma(u)) &= \sup_{u\in U}\sup_{x\in X} \nu_j(\Sigma(x)(u)) \\ &= \sup_{x\in X} \omega^x_{U,j}(\Sigma(x))\end{aligned}$$

is finite. Clearly, this implies the continuity of T_Σ.

Case b. By [Sch 71, II.8.3] we have to show that $(T_\Sigma(u_n))_{n\in\mathbb{N}}$ converges to 0 for every 0-sequence $(u_n)_{n\in\mathbb{N}}$ in F.
Fix $\varepsilon > 0$ and let $\nu_j : E \to \mathbb{R}$ be a seminorm of the bundle $p : E \to X$.
If $(u_n)_{n\in\mathbb{N}}$ is a fixed 0-sequence in F, we show :

(**) For every $x \in X$ there is a neighborhood U of x and a natural number $N \in \mathbb{N}$ such that for all $n \geq N$ and all $y \in U$ we have $\sup_{y\in U} \nu_j(T_\Sigma(u_n)(y)) \leq \varepsilon$.

Once (**) is established, an easy compactness argument will finish the proof.
To convince the reader of (**), we shall again use (*): Firstly, choose again any $T \in L$ such that $\varepsilon_x \circ T = \Sigma(x)$ and let $S = \{0\} \cup \{u_n : n \in \mathbb{N}\}$. Note that S belongs to S by our assumption. Thus (*) yields an open neighborhood U of x such that

$$\sup_{y\in U} \nu_j(T_\Sigma(u_n)(y) - T(u_n)(y)) \leq \varepsilon/2$$

for all $n \in \mathbb{N}$. As the operator $T : F \to \Gamma(p)$ is continuous, we conclude that $\lim_{n\to\infty} T(u_n) = 0$. Therefore there is an $N \in \mathbb{N}$ such that

$$\sup_{y\in X} \nu_j(T(u_n)(y)) \leq \varepsilon/2$$

for all $n \in \mathbb{N}$. Using the triangle inequality, these two inequalities together yield (**).

To show that the mapping $\Sigma \to T_\Sigma$ is an embedding, we have to recall that the topology on $L_S(F,\Gamma(p))$ is induced by the seminorms $\vartheta_{S,j}$, $j \in J$, $S \in \mathcal{S}$ given by

$$\vartheta_{S,j}(T) = \sup_{u\in S} \sup_{x\in X} \nu_j(T(u)(x))$$

and the topology on $\Gamma(q_L)$ is given be the seminorms $\omega_{S,j}$, $j \in J$ and $S \in \mathcal{S}$ defined by

$$\omega_{S,j}(\Sigma) = \sup_{x\in X} \sup_{u\in S} \nu_j(\Sigma(x)(u)).$$

An easy computation shows that for $\Sigma \in \Gamma(q_L)$ we have

$$\omega_{S,j}(\Sigma) = \vartheta_{S,j}(T_\Sigma)$$

and thus the proof is complete. □

11.12 Corollary. *Let* $p : E \to X$ *be a bundle with a completely regular base space and let* F *be a topological vector space. Then there is a bundle* $q : F \to X$ *such that* $L_{pc}(F,\Gamma(p))$ *equipped with the topology of precompact convergence is isomorphic to a* $C_b(X)$*-submodule of* $\Gamma(q)$. *The stalk over* $x \in X$ *of this bundle may be choosen to be a subspace of* $L_{pc}(F,E_x)$, *where* E_x *is the stalk over* x *of the bundle* $p : E \to X$. *In this case, the canonical injection* $\Phi : L_{pc}(F,\Gamma(p)) \to \Gamma(q)$ *is given as in* (11.9)
Moreover, in each of the following cases (a) *and* (b), *the map* Φ *is surjective with inverse*

$$\Psi : \Gamma(q) \to L_{pc}(F,\Gamma(p))$$
$$\Sigma \to T_\Sigma$$

where $T_\Sigma(u)(x) = \Sigma(x)(u)$ *for all* $u \in F$ *and all* $x \in X$*:*

a) F is finite dimensional.

b) F is bornological and X is compact.

Proof. Only the verification of the surjectivity of Φ is of some interest. But this follows from (11.10), if we note that $0 \in F$ has a precompact neighborhood, provided that F is finite dimensional, whence case (a) of (11.10(ii)) applies under these circumstances. Moreover, the closure of every 0-sequence is precompact and thus case (b) of (11.10(ii)) applies in case (b) of (11.12). □

Our next corollary concerns spaces of compact operators. Recall that an operator $K : F \to E$ between topological vector spaces is called compact, if there is a neighborhood U of $0 \in F$ such that K(U) is relatively compact in E. By $\mathcal{K}(F,E)$ we denote the subspace of $\mathcal{L}_b(F,E)$ of all compact operators, equipped with the topology of bounded convergence. If $U \subset F$ is a neighborhood of 0, we let

$$\mathcal{K}_U(F,E) = \{K \in \mathcal{K}(E,F) : K(U) \text{ is relatively compact}\}$$

and we equip this space with the topology of uniform convergence on U (which may be finer than the topology inherited from $\mathcal{K}(F,E)$

If $p : E \to X$ is a bundle, then $\mathcal{K}(F,\Gamma(p))$ and $\mathcal{K}_U(F,\Gamma(p))$ are $C_b(X)$-submodules of $\mathcal{L}(F,\Gamma(p))$. Therefore, we can state:

11.13 Corollary. *Let* $p : E \to X$ *be a bundle with a completely regular base space, let* F *be a topological vector space and let* $U \subset F$ *be a neighborhood of* 0*. Then there is a bundle* $q : F \to X$ *such that* $\mathcal{K}_U(F,\Gamma(p))$ *is isomorphic to a* $C_b(X)$*-submodule of* $\Gamma(q)$*. The stalk over* $x \in X$ *of this bundle may be choosen to be a subspace of*

$K_U(F,E_x)$, where $E_x = p^{-1}(x)$. In this case, the canonical injection $\Phi : K_U(F,\Gamma(p)) \to \Gamma(q)$ is given as in (11.9). Moreover, we have a (topological) embedding

$$\begin{array}{cccc} \Psi : & \Gamma(q) & \to & L_U(F,\Gamma(p)) \\ & \Sigma & \to & T_\Sigma \end{array}$$

where $L_U(F,\Gamma(p))$ denotes the space $L(F,\Gamma(p))$ equipped with the topology of uniform convergence on U.
If X is compact and if all stalks of the bundle $p : E \to X$ are quasicomplete, then Φ is a bijection with inverse Ψ.

Proof. We again apply (11.9) to establish the existence of such a bundle. Note that the stalk F_x of the bundle $q : F \to X$ may be identified with $\{\varepsilon_x \circ K : K \in K_U(F,\Gamma(p))\}$ and hence is contained in $K_U(F,E_x)$. The fact that $\Psi : \Gamma(q) \to L_U(F,\Gamma(p))$ is a topological embedding follows from (11.10(ii)), case (a).

Finally, if X is compact, then the image of $K_U(F, \Gamma(p))$ under Φ is dense in $\Gamma(q)$ by the Stone-Weierstraß theorem (4.2). As the restriction of Ψ to the image of Φ is the inverse of Φ, this implies that $K_U(F,\Gamma(p))$ is dense in the image of Ψ. Now we know from (1.10) that $\Gamma(p)$ is quasicomplete whenever all the stalks are quasicomplete. From the proof of (III.9.3) in [Sch 71] we conclude that $K_U(F,\Gamma(p))$ is closed in $L_U(F,\Gamma(p))$. This shows that $K_U(F,\Gamma(p))$ is equal to the image of Ψ and Ψ is the inverse of Φ. □

Of course, we can apply (11.13) to $K(F,\Gamma(p))$, where F is a normed space. If in addition $p : E \to X$ is a bundle of Banach spaces, then we obtain:

11.14 Corollary. *Let $p : E \to X$ be a bundle of Banach spaces,*

X completely regular, and let F be a normed space. Then there is a bundle $q : \mathcal{F} \to X$ of Banach spaces such that the Banach space $K(F,\Gamma(p))$ of all compact operators equipped with the operator norm is isometrically isomorphic to a $C_b(X)$-submodule of $\Gamma(q)$. The stalk over $x \in X$ of this bundle may be choosen to be a closed subspace of $K(F,E_x)$, equipped with the operator norm. In this case, the canonical injection $\Phi : K(F,\Gamma(p)) \to \Gamma(q)$ is given as in (11.9).
If X is compact, then Φ is bijective.

Proof. This result is a variation of (11.13); there are two things which have to be checked:

(i) The mapping Φ is an isometry: This follows immediately from the definition of the operator norm, the definition of the stalks of the bundle $q : \mathcal{F} \to X$ as it was given in the proof of (11.9) and (11.8(ii)).

(ii) The stalks, as they have been defined in the proof of (11.9), are Banach spaces and thus closed subspaces of $K(F,E_x)$, $x \in X$ From (11.7) we may conclude that the stalks are isometrically isomorphic to quotients of $K(F,\Gamma(p))$ and thus are complete, since $K(F,\Gamma(p))$ is a Banach space. □

In these last three corollaries the stalks of the bundle $q : E \to X$ were always subspaces of larger spaces: They were subspaces of $L_{pc}(F,E_x)$ in (11.12), subspaces of $K_U(F,E_x)$ in (11.13) and subspaces of $K(F,\Gamma(p))$ in (11.14). In which cases do we obtain the whole space as stalk? It turns out that at least in the first and in the last case the answers are the same: It suffices that all stalks of the bundle $p : E \to X$ have the approximation property in the sense of Grothendieck (see [Gr 55]). Alternatively, we could postulate that the bundle $p : E \to X$ is locally trivial.

The problem we are dealing with in this context is the following: Given a point $x \in X$ in the base space of the bundle $p : E \to X$ and an operator $t : F \to E_x$, can we find a "lifting" $T : F \to \Gamma(p)$ such that $\varepsilon_x \circ T = t$?

11.15 Proposition. *Let* $p : E \to X$ *be a bundle over a completely regular base space, let* F *be a topological vector space and let* S *be a directed and total family of bounded subsets of* F*. Then for every* $x \in X$*, the closure of* $\{\varepsilon_x \circ T : T \in L_S(F,\Gamma(p))$ and $\dim T(F) < \infty\}$ *in* $L_S(F,E_x)$ *contains all operators of finite rank.*

Proof. Let $t \in L_S(F,E_x)$ be of finite rank, i.e.

$$t = \sum_{i=1}^{n} \phi_i \otimes \alpha_i$$

for certain elements $\alpha_i \in E_x$ and certain elements $\phi_i \in F'$. Given $S \in S$ and an open, convex and circled neighborhood $U \subset E_x$ of 0, we have to find an element $T \in L_S(F,\Gamma(p))$ such that $\dim T(F) < \infty$ and such that $(t - \varepsilon_x \circ T)(S) \subset U$.
Firstly, note that $\phi_i(S)$ is bounded in $\mathbb{K}$ for every $1 \le i \le n$. Thus, we can find a constant $M > 0$ such that $|\phi_i(s)| \le M$ for all $s \in S$ and all $i \in \{1,\dots,n\}$. Moreover, by (1.5.III) and (2.2), the set $\{\sigma(x) : \sigma \in \Gamma(p)\} \subset E_x$ is dense in E_x. Hence we can find sections $\sigma_1,\dots,\sigma_n \in \Gamma(p)$ such that $\alpha_i - \sigma_i(x) \in \frac{1}{M\cdot n}\cdot U$ for all $i \in \{1,\dots,n\}$.
Now define

$$T := \sum_{i=1}^{n} \phi_i \otimes \sigma_i \quad : \quad F \to \Gamma(p).$$

Then, by definition, T is of finite rank and for all $s \in S$ we have

$$\begin{aligned}(t - \varepsilon_x \circ T)(s) &= \Big(\sum_{i=1}^{n} \phi_i \otimes (\alpha_i - \sigma_i(x))\Big)(s) \\ &= \sum_{i=1}^{n} \phi_i(s)\cdot(\alpha_i - \sigma_i(x))\end{aligned}$$

$$\in \sum_{i=1}^{n} \phi_i(s) \cdot \frac{1}{M \cdot n} \cdot U$$

$$\subseteq U ,$$

i.e. $(t - \varepsilon_x \circ T)(S) \subset U$. □

It is now evident that we are lead to spaces with the approximation property:

11.16 Definition. A locally convex topological vector space E has the *approximation property*, provided that for every locally convex topological vector space F the linear operators of finite rank from E into F are dense in $L_{pc}(F,E)$. □

A. Grothendieck ([Gr 55]) showed that for Banach spaces E this definition is equivalent to the following statement:
For every normed space F the linear operators of finite rank from E into F are dense in $K(F,E)$.

We now can state:

11.17 Complement. *(i) Let* $p : E \to X$ *be a bundle over a completely regular base space* X *such that all the stalks have the approximation property. Then the stalks of the bundle* $q : F \to X$ *in* (11.12) *may be chosen to be dense subspaces of* $L_{pc}(F,E_x)$, $x \in X$.
(ii) If in addition $p : E \to X$ *is a bundle of Banach spaces, then the stalks of the bundle* $q : F \to X$ *in* (18.14) *may be chosen to be* $K(F,E_x)$, $x \in X$. □

11.18 Remark. Under the conditions of (11.12) and (11.17) we can choose the whole spaces $L_{pc}(F,E_x)$, $x \in X$, as the stalks of the bundle

$q : F \to X$. In this case however, it may happen that the bundle $q : F \to X$ is no longer a full bundle, although I do not know of any example to illustrate this.

With this new choice of the stalks even the second half of (11.12) remains valid. To show this, we would have to generalize (11.10), notabely the properties (*) and (**) in the proof of (11.10). As we are not going to use these facts in the following, we leave the details to the reader.

In the next theorem we apply the results obtained so far to the approximation property of spaces of sections:

11.19 Theorem. *Let* $p : E \to X$ *be a bundle over a compact base space* X. *Then the space of all sections* $\Gamma(p)$ *has the approximation property, provided that every stalk* E_x, $x \in X$, *has the approximation property.*

Proof. Let F be a topological vector space and let $F' \otimes \Gamma(p)$ be the set of all linear operators from F into $\Gamma(p)$ of finite rank. We have to show that $F' \otimes \Gamma(p)$ is dense in $L_{pc}(F, \Gamma(p))$.

Firstly, note that $F' \otimes \Gamma(p)$ is a C(X)-submodule of $L_{pc}(F, \Gamma(p))$, since the multiplication with elements $f \in C(X)$ is linear. From (11.12) and (11.17(i)) we know that there is a bundle $q : F \to X$ with stalks isomorphic to the dense subspaces $\{\varepsilon_x \circ T : T \in L_{pc}(F, \Gamma(p))\}$ of $L_{pc}(F, E_x)$, $x \in X$, such that $L_{pc}(F, \Gamma(p))$ may be identified with a C(X)-submodule of $\Gamma(q)$. Under these identifications the set $\{T(x) : T \in F' \otimes \Gamma(p)\}$ is dense in $\{\varepsilon_x \circ T : T \in L_{pc}(F, \Gamma(p))\}$ by (11.15). Hence the Stone-Weierstraß theorem (4.2) yields that $F' \otimes \Gamma(p)$ is dense in $L_{pc}(F, \Gamma(p))$. □

For a more detailed discussion of the approximation property of spaces of sections, we refer to [Gi 78],[Pr 79], and [Bi 80].

Another important case of $C(X)$-submodules of $L(F,\Gamma(p))$ was already discussed in section 10 and we shall add some facts here:

Let us consider a second bundle $p' : E' \to X$. Then the set of all $C_b(X)$-module homomorphisms from $\Gamma(p')$ into $\Gamma(p)$ form a $C_b(X)$-submodule of $L(\Gamma(p'),\Gamma(p))$. We shall assume that $p' : E' \to X$ is a full bundle and that the base space X is completely regular. Under these conditions we saw in (10.7) that every continuous $C_b(X)$-module homomorphism $T : \Gamma(p') \to \Gamma(p)$ may be "decomposed" into a bundle morphism $\lambda_T : E' \to E$ and this "decomposition" may be indeed been thought of as a section in the bundle constructed in (11.9). To explain this, let us start with a lemma:

11.20 Lemma. *Let* E *and* F *be locally convex topological vector space, let* M *be a closed subspace of* F *and let* $\pi : F \to F/M$ *be the quotient map. If* S *is an updirected and total family of bounded subsets of* F, *then the mapping*

$$\iota_\pi : L_{\pi(S)}(F/M,E) \to L_S(F,E)$$
$$T \to T\circ\pi$$

is a topological embedding with range $\{T \in L_S(F,E) : T(M) = 0\}$.

Proof. Let $T \in L_S(F,E)$ and assume that $T(M) = 0$. Then we have $T(S) \subset U$ if and only if $T(S + M) \subset U$, where $S \in S$ and where $U \in E$ is an open neigborhood of 0. □

Let us apply (11.9) to the situation where $F = \Gamma(p')$ for a full bundle $p' : E' \to X$, where S is a directed family of precompact subsets of $\Gamma(p')$ such that F is generated by $\cup\, \mathcal{S}$ and where $\mathcal{L} = \mathrm{Mod}(\Gamma(p'),\Gamma(p))$. Then we find a bundle $q : \mathcal{F} \to X$ such that $\mathcal{L}$ is (isomorphic to) a $C_b(X)$-submodule of $\Gamma(q)$, the stalks of this bundle being $\{\varepsilon_x \circ T : T \in \mathrm{Mod}(\Gamma(p'),\Gamma(p))\} \subset \mathcal{L}_{\mathcal{S}}(\Gamma(p'),E_x)$. As $p' : \mathcal{E}' \to X$ is a full bundle, the evaluation map $\varepsilon_x : \Gamma(p') \to p'^{-1}(x)$ is a (topological) quotient map by (2.7). Hence by (10.6) and (11.20) the subspace $\{\varepsilon_x \circ T : T \in \mathrm{Mod}(\Gamma(p'),\Gamma(p))\} \subset \mathcal{L}_{\mathcal{S}}(\Gamma(p'),E_x)$ may be identified with a subspace of $\mathcal{L}_{\mathcal{S}(x)}(E_x',E_x)$, where $E_x' = p'^{-1}(x)$ and where $\mathcal{S}(x) = \{\varepsilon_x(S) : S \in \mathcal{S}\}$. Under this identification, the operator $\varepsilon_x \circ T : \Gamma(p') \to E_x$ corresponds to the unique operator $T_x : E_x' \to E_x$ such that the diagram

$$\begin{array}{ccc} \Gamma(p') & \xrightarrow{T} & \Gamma(p) \\ \varepsilon_x \downarrow & & \downarrow \varepsilon_x \\ E_x' & \xrightarrow[T_x]{} & E_x \end{array}$$

is commutative. It is clear from the proof of (10.5) and (10.7) that $T_x = \lambda_T | p'^{-1}(x)$, where $\lambda_T : \mathcal{E}' \to \mathcal{E}$ is the unique bundle morphism such that $T = T_\lambda$. Let us agree that we write $\lambda_T(x)$ instead of $\lambda_{T/p'^{-1}(x)}$.

Applying (11.9) we obtain a bundle $q : \mathcal{F} \to X$ such that $\mathrm{Mod}(\Gamma(p'),\Gamma(p)) \subset \mathcal{L}_{\mathcal{S}}(\Gamma(p'),\Gamma(p))$ is isomorphic to a $C_b(X)$-submodule of $\Gamma(q)$. The stalks of this bundle may be chosen to be subspaces of $\mathcal{L}_{\mathcal{S}(x)}(E_x',E_x)$ and the canonical injection is given by $\lambda \to \lambda_T$. Furthermore, the family of seminorms of the bundle $q : \mathcal{F} \to X$ is defined by

$$\begin{aligned} \omega_{S,j} \;:\; & \mathcal{F} \to \mathbb{R} \\ & \Lambda \to \sup_{s \in S} \nu_j(\Lambda\{s[q(\Lambda)]\}), \end{aligned}$$

where $S \in \mathcal{S}$ and where $\nu_j : \mathcal{E} \to \mathbb{R}$ is one of the seminorms of the bundle $p : E \to X$.

If X is compact and if $\Gamma(p')$ is bornological, then $\mathrm{Mod}(\Gamma(p'),\Gamma(p))$ and $\Gamma(q)$ are isomorphic, provided that S contains the closure of every 0-sequence.

We state a special case of these observations as a theorem:

11.21 Theorem. *Let* $p : E \to X$ *and* $p' : E' \to X$ *be bundles of Banach spaces over a compact base space* X. *Then there is a bundle* $q : F \to X$ *such that the* $C(X)$*-module* $\mathrm{Mod}(\Gamma(p'),\Gamma(p))$ *equipped with the topology of compact convergence is topologically and algebraically isomorphic to* $\Gamma(q)$. *The stalks of this bundle may be chosen to be subspaces of* $L_c(E_x',E_x)$. *In this case, the canonical isomorphism is given by*

$$\lambda_- : \mathrm{Mod}(\Gamma(p'),\Gamma(p)) \to \Gamma(q)$$
$$T \to \lambda_T$$

Proof. Let S denote the family of all compact subsets of $\Gamma(p')$. If we can show that $S(x)$ is the family of all compact subsets of E_x', the theorem will follow from the discussions preceeding (11.21).

Thus, we are dealing with the following problem: Given a Banach space E, a closed linear subspace F, a compact subset $A \subset E/F$, is there a compact subset $B \subset E$ such that $A = B + F$?. But this is a well-known result from the theory of Banach spaces. □

12. Excursion: Continuous lattices and bundles

In the past years a certain type of lattices appeared in mathematics, which seem to be a natural background of a large variety of order theoretical properties of mathematical structures. These lattices were called continuous lattices by D.Scott in [Sc 72]. In the following years, K.H.Hofmann and A.Stralka discovered that this type of lattices was already known to other mathematicians in different areas. J.D.Lawson, for instance, called them compact topological semilattices with small semilattices, moreover A.Day and O.Wyler found them as "algebras" of the filter monad in category theory.

Also in functional analysis the concept of continuous lattices seems to be useful. In this section we shall collect a few results and definitions which will be needed later on. With a few exceptions, the proofs may be found in [Comp 80].

12.1 Let L be a complete lattice. A subset $D \subset L$ is said to be *directed*, if every pair $a, b \in D$ has an upper bound in D.
If $a, b \in L$ are two elements, we say that a is *way below* b, if every directed set D with $\sup D \geq b$ contains an element $d \in D$ such that $a \leq d$.

We shall abbreviate the phrase "a is way below b" by writing $a \ll b$.

12.2 A complete lattice L is called *continuous lattice*, if for all $a \in L$ we have $a = \sup \{b : b \ll a\}$.

12.3 We add a couple of examples which will be of significance:

(i) Let X be a locally compact topological space. By $O(X)$ we denote the complete lattice of all open subsets of X, ordered by inclusion. Then $O(X)$ is a continuous lattice. Moreover, we have $U \ll V$ if and only if $\overline{U}$ is compact and contained in V (i.e. if U is relatively compact in V in the topological sense).

(ii) Let K be a compact convex subset of a locally convex topological vector space and let Conv(K) be the complete lattice of all closed convex subsets of K, ordered by dual inclusion (i.e. $A \leq B$ iff $B \subset A$). Then Conv(K) is a continuous lattice. Here, we have

$$A \vee B = A \cap B$$

$$A \wedge B = \overline{\mathrm{conv}}(A \cup B)$$ (where $\overline{\mathrm{conv}}(M)$ denotes the closed convex hull of M)

$$A \ll B \text{ iff } B \subset A^{o}$$ (where ^{o} is the topological kernel operator)

12.4 Every continuous lattices carries two important topologies, which we will use later on:

(i) *The Scott topology*. Let L be a (continuous) lattice. A subset $U \subset L$ is said to be *Scott-open* if

(1) $u \in U$ and $u \leq v$ imply $v \in U$

(2) If $D \subset L$ is directed and if $\sup D \in U$, then $U \cap D \neq \emptyset$.

It is easy to verify that the Scott open sets form a topology on L which will be called the *Scott-topology*.

In a continuous lattice the sets of the form $V(a) := \{x \in L : a \ll x\}$ form a base for the Scott-topology. A mapping $f : L \to V$ between two complete lattices is *Scott-continuous* (i.e. is continuous with respect to the Scott-topologies on L and V) if and only if for every directed subset $D \subset L$ we have $f(\sup D) = \sup f(D)$.

(ii) The topology generated by the Scott-topology together with all sets of the form $L \setminus \uparrow a$, $a \in L$, is called the *Lawson-topology*. On a continuous lattice, the Lawson-topology is always compact and Hausdorff. Further, the mapping $\wedge : L \times L \to L$ is continuous and the Lawson topology is uniquely determined by these properties.

An $\wedge$-homomorphism $f : L \to V$ between continuous lattices L and V is Lawson-continuous (i.e. continuous with respect to the Lawson topologies) if and only if f preserves suprema of directed sets and arbitrary infima.

12.5 If X is locally compact and if $A \subset X$ is a compact subset of X, then $\{U \in O(X) : A \subset U\}$ is a typical Scott-open set. Of course, instead of using O(X) we may consider the complete lattice Cl(X) of all closed subsets of X, ordered by dual inclusion. In this case, $\{B \in Cl(X) : A \cap B \neq \emptyset\}$ is Scott-open for every compact subset $A \subset X$.

Let K be a compact convex subset of a locally convex topological vector space. If $U \subset K$ is relatively open in K, then $\{A \in Conv(K) : A \subset U\}$ is Scott open in Conv(K).

12.6 Let X again be locally compact. Then the Lawson topology on O(X) (or, equivalently, on Cl(X)) is the well-known Hausdorff topology. We will see in a moment that the same is true for the continuous lattice Conv(K);

12.7 Proposition. *If* K *is a compact convex set in a locally convex topological vector space, then the inclusion* Conv(K) $\to$ Cl(X) *is continuous for the resp. Lawson topologies. Especially,* Conv(K) *is closed in* Cl(K).

Proof. Firstly, we show that Conv(K) is closed in Cl(K). Let $A \in Cl(K) \setminus Conv(K)$ be a closed subset of K which is not convex. We have to find an open neighborhood of A which does not intersect Conv(K). Pick a $\lambda \in [0,1]$ and elements $a,b \in A$ such that $\lambda \cdot a + (1-\lambda) \cdot b =: c \notin A$. Let W be an open set around c such that $\overline{W} \cap A = \emptyset$. As the mapping $(x,y) \to \lambda \cdot x + (1-\lambda) \cdot y : K \times K \to K$ is continuous, there are open sets U,V around a and b resp. such that $\lambda \cdot U + (1-\lambda) \cdot V \subset W$. Now the set

$$\{C \in Cl(K) : C \not\subset K \setminus U,\ C \not\subset K \setminus V,\ C \cap \overline{W} = \emptyset\}$$

is open in the Lawson topology of Cl(K) and contains A. Moreover, this open set is disjoint from Conv(K) : If $C \not\subset K \setminus U$, $C \not\subset K \setminus V$ and $C \cap \overline{W} = \emptyset$, we may pick elements $x \in C \cap U$ and $y \in C \cap V$. Then the convex combination $\lambda \cdot x + (1-\lambda) \cdot y$ belongs to $\lambda \cdot U + (1-\lambda) \cdot V \subset$ $\subset W$ and therefore cannot belong to C as $C \cap X = \emptyset$. Hence C is not convex.

Next, we claim that the Lawson topology on Conv(K) is coarser than the topology induced by the Lawson topology on Cl(K). This will finish the proof, as both topologies are compact.

Let $\mathcal{U} \subset Conv(K)$ be Scott open and let $A \in \mathcal{U}$. By (12.3) and (12.4) we may find a $B \in Conv(K)$ such that $A \subset B^{\circ}$ and such that $C \subset B^{\circ}$ implies $C \in \mathcal{U}$ for all $C \in Conv(K)$. The set $\mathcal{S} := \{C \in Cl(K) : C \subset B^{\circ}\}$ is open in Cl(K) and we have $A \in \mathcal{S} \cap Conv(K) \subset \mathcal{U}$. Hence $\mathcal{U}$ is open in the topology induced by the Lawson topology on Cl(K).

Finally, let $A \in Conv(K)$ and let $\mathcal{V} = \{B \in Conv(K) : B \nsupseteq A\} =$ $= \{B \in Conv(K) : A \not\subset B\}$. Then $\mathcal{V} = Conv(K) \cap \{B \in Cl(K) : A \not\subset B\}$ and therefore $\mathcal{V}$ is open in the topology induced by Cl(K), too. As these two types of sets generate the Lawson topology on Conv(K), our proof is complete. □

12.8 Let us return to continuous lattices in general. If L is a continuous lattice and if $\mathcal{U}$ is an ultrafilter on L, we know that $\mathcal{U}$ has to converge in the Lawson topology on L. The limit of this ultrafilter may by calculated as follows:

$$\lim \mathcal{U} = \sup_{M \in \mathcal{U}} \inf M \quad .$$

Translated to converging nets $(x_i)_{i \in I}$ in L, the formula reads as

$$\lim_{i \in I} x_i = \sup_{i \in I} \inf_{j \geq i} x_j \quad .$$

12.9 An element $p \in L$ of a lattice L is called *prime*, if $a \wedge b \leq p$ implies $a \leq p$ or $b \leq p$ for all $a,b \in L$.
Prime elements in a continuous lattice have a much stronger property, as the following lemma shows:

12.10 Lemma *Let* L *be a continuous lattice, let* V *be a complete lattice and let* $f : L \to V$ *be a mapping such that* $f(\sup D) = \sup f(D)$ *for every directed set* $D \subset L$. *If* $A \subset L$ *is compact in the Lawson-topology and if* $p \in V$ *is a prime element of* V, *then* $\inf f(A) \leq p$ *implies* $f(a) \leq p$ *for some* $a \in A$. □

For a proof of (12.10) we refer to [GK 77] or [Comp 80].

Let us now return to bundles. One connection between bundles and continuous lattices comes out of the following considerations:

12.11 Suppose that $p : E \to X$ is a bundle with a compact base space. Then $O(X)$ is a continuous lattice. Moreover, we have a canonical mapping between $O(X)$ and the complete lattice $C(\Gamma(p))$ of all closed subspaces of $\Gamma(p)$ given by

$$i : O(X) \to C(\Gamma(p))$$
$$U \to N_{X\setminus U} = \{\sigma \in \Gamma(p) : \sigma_{/X\,U} = 0\}\ .$$

This mapping satisfies the hypothesis of (12.10) (see also [GK 77]):

12.12 Proposition. *Let* $p : E \to X$ *be a bundle with a compact base space. Then the mapping* $i : O(X) \to C(\Gamma(p))$ *preserves directed suprema, i.e. if* $(U_\lambda)_{\lambda\in\Lambda}$ *is a directed family of open subsets of* X*, then*

$$i(\bigcup_{\lambda\in\Lambda} U_\lambda) = (\bigcup_{\lambda\in\Lambda} i(U_\lambda))^-$$

Proof. The mapping i is monotone, whence we have the inclusion

$$i(\bigcup_{\lambda\in\Lambda} U_\lambda) \supset (\bigcup_{\lambda\in\Lambda} i(U_\lambda))^-.$$

Conversely, let $\sigma \in i(\bigcup_{\lambda\in\Lambda} U_\lambda)$. We have to show: For every $\varepsilon > 0$ and every seminorm ν_j belonging to the bundle there is an $\sigma' \in \bigcup_{\lambda\in\Lambda} i(U_\lambda)$ such that $\hat{\nu}_j(\sigma - \sigma') \le \varepsilon$.
Thus, let $\varepsilon > 0$ and ν_j be given. We define $U := \{x \in X : \nu_j(\sigma(x)) < \varepsilon\}$. As $\sigma_{|X\setminus U} = 0$, we obtain $X \setminus \bigcup_{\lambda\in\Lambda} U_\lambda = \bigcap_{\lambda\in\Lambda} (X \setminus U_\lambda) \subset U$. Now the compactness of X and the fact that $(U_\lambda)_{\lambda\in\Lambda}$ is directed yields a $\lambda_o \in \Lambda$ such that $X \setminus U_{\lambda_o} \subset U$. Choose any continuous function $f : X \to [0,1] \subset \mathbb{R}$ such that $f(X \setminus U_{\lambda_o}) = \{0\}$and $f(X \setminus U) = \{1\}$ and set $\sigma' = f\cdot\sigma$. Obviously, the section σ' belongs to $i(U_{\lambda_o}) \subset$ $\subset \bigcup_{\lambda\in\Lambda} i(U_\lambda)$. An easy calculation shows that $\hat{\nu}_j(\sigma - \sigma') \le \varepsilon$. □

If we take (12.10) and (12.12) together, we have done most of the proof of the following

12.13 Proposition. *Let* $p : E \to X$ *be a bundle over a compact base space and let* $V \subset C(\Gamma(p))$ *be any complete lattice of closed subspaces of* $\Gamma(p)$ *containing all subspaces of the form*

$N_A = \{\sigma \in \Gamma(p) : \sigma_{/A} = 0\}$, A *closed. If* $P \neq \Gamma(p)$ *is a prime element of* V, *the there is a unique* $x \in X$ *such that* $N_x \subset P$.

Proof. Suppose that we would have $N_x \subset P$ and $N_y \subset P$, where $x \neq y$. Let $f : X \to \mathbb{K}$ be any continuous function which takes the value 0 at x and the value 1 at y. If $\sigma \in \Gamma(p)$, then we may write

$$\begin{aligned} \sigma &= f\cdot\sigma + (1 - f)\cdot\sigma \\ &\in N_x + N_y \\ &\subset P, \end{aligned}$$

and hence $P = \Gamma(p)$, a contradiction. This shows that there is at most one such x.

To ensure the existence of such $x \in X$, we let

$$K := \{X \setminus \{x\} : x \in X\} \subset O(X).$$

It is well known that K is compact in the Lawson-topology of $O(X)$ (see [Comp 80]). Moreover, $\bigcap i(K) = \bigcap_{x\in X} N_x = \{0\}$. Hence, if P is prime in V, we can find an $x \in X$ such that $N_x \subset P$ by (12.10) and (12.12). □

12.14 Let L be any complete lattice. By Spec(L) we abbreviate the set of all prime elements of L which are different from the largest element 1. The sets of the form

$$s(a) := \{p \in \text{Spec}(L) : a \not\leq p\} \quad , a \in L,$$

form a topology on Spec(L), the so called *hull-kernel-topology*. The closed sets of this topology are exactly the sets of the form

$$h(a) := \{p \in \text{Spec}(L) : a \leq p\} \quad , a \in L.$$

12.15 If $p : E \to X$ is a bundle with a compact base space and if

$V \subset C(\Gamma(p))$ is a complete lattice of closed subspaces containing $\{N_A : A \in Cl(X)\}$, then (12.13) means that we have a mapping

$$\text{fix} : \text{Spec}(V) \to X$$

which sends every prime element $P \in \text{Spec}(V)$ to the unique $x \in X$ such that $N_x \subset P$. (In this case we say: P *is fixed at* x.)

12.16 Proposition. *If* $p : E \to X$ *is our favorite bundle with a compact base space and if* $V \subset C(\Gamma(p))$ *is a complete lattice of closed subspaces containing* $\{N_A : A \in Cl(X)\}$, *then the mapping*

$$\text{fix} : \text{Spec}(V) \to X$$

is continuous, where Spec(V) *carries the hull-kernel-topology.*

Proof. Let $A \subset X$ be closed. With the notations of (12.14) we show that $\text{fix}^{-1}(A) = h(N_A)$.
Indeed, if $\text{fix}(P) \in A$, then $N_A \subset N_{\text{fix}(P)} \subset P$, i.e. $P \in h(N_A)$.
Conversely, assume that $P \in h(N_A)$ but $x_o := \text{fix}(P) \notin A$. Then we obtain $N_A \subseteq P$ and $N_{x_o} \subset P$. Using Uryson's lemma, we show as in the proof of (12.13) that $\Gamma(p) = N_{x_o} + N_A \subset P$, contradicting $P \neq \Gamma(p)$. □

13. M-structure and bundles

Let E be a C(X)-Ω-module. Again, we ask: Under which conditions is it true that E is isomorphic to a space of sections in a bundle? We saw in (7.21) and (7.23) that for the algebraic point of view the Banach algebra C(X), viewed as a space of bounded operators on E, should be contained in the "center" of the Ω-space E. It is a bit surprising that on the topological side there is also a notion of center of E. We shall see that for Banach spaces E there is a commutative, closed subalgebra $Z_t(E) \subset \mathcal{B}(E)$ containing the identity such that E is locally C(X)-convex if and only if $C(X) \subset Z_t(E)$. Hence, if we are interested in representations by sections in a bundle, the intersection of the two "centers" $Z_t(E)$ and $Z_\Omega(E)$ is the right object to look at. We shall see furthermore, that the intersection $Z_t(E) \cap Z_\Omega(E)$ is a commutative Banach algebra with unit and thus of the form C(X) for a certain compact Hausdorff space X. This space X is the maximal space such that the Ω-space E may be represented as the space of all sections in a bundle $p : \mathcal{E} \to X$ of Ω-spaces. Unfortunately, it is difficult to get hold of the stalks of this bundle. They are by no means "indecomposable" in the sense that $Z_t(E_x) \cap Z_\Omega(E_x)$ is one-dimensional, or, in other words, that every bundle representation of the stalk with a compact base space leads to a one-point base space.

In the same sense as the topological center $Z_t(E)$ is the counterpart of the algebraical center, we shall find a topological analog of Ω-ideals: The M-ideals. It is remarkable that, as in the algebraical situation, the topological center $Z_t(E)$ may be represented as the Banach algebra of all continuous $\mathbb{K}$-valued functions on the space of all "primitive" M-ideals, equipped with the hull-kernel topology.

In this section, we shall restrict ourselves to Banach spaces and bundles of Banach spaces.

Let us start with a list of results from the theory of M-structure in Banach spaces. The proofs may all be found in the lecture notes of E.Behrends ([Be 79]) or in the earlier paper of E.M.Alfsen and E.G.Effros ([AE 72]).

13.1 Let E be a real or complex Banach space. A *projection* on E is a continuous and linear map $p : E \to E$ such that $p \circ p = p$. A projection p is called an L-*projection*, if we have in addition

$$\|m\| = \|p(m)\| + \|m - p(m)\| \quad \text{for all } m \in E.$$

If p,q are two L-projections on the same Banach space E, then $p \circ q = q \circ p$, i.e. L-projections commute. Moreover, the operators

$$\begin{aligned} p \wedge q &:= p \circ q \\ p \vee q &:= p + q - p \circ q \\ p^{\perp} &:= \mathrm{Id} - p \end{aligned}$$

are L-projections, too. Hence the L-projections form a Boolean algebra. In this Boolean algebra, we have $p \le q$ iff $p \circ q = p$. Moreover, if $(p_i)_{i \in I}$ is an increasing family of L-projections, then this family is pointwise convergent to an L-projection. Thus, the Boolean algebra of all L-projections is complete and we denote this Boolean algebra by $\mathbb{P}_L(E)$.

13.2 The Banach algebra Cu(E) generated by $\mathbb{P}_L(E)$ in $\mathcal{B}(E)$ is called the *Cunningham algebra of* E. As L-projections commute, Cu(E) is a commutative algebra with unit, which is of the form $C(X)$, where X is a compact space. Moreover, the space X may be identified with the Stone dual $\mathbb{P}_L(E)^\wedge$ of the Boolean algebra $\mathbb{P}_L(E)$ and thus is an extremaly disconnected space. This implies that the Cunningham algebra is an order complete Banach lattice, i.e. if $M \subset Cu(E)$ is order bounded, then sup M exists in $Cu(E)$. If M is in addition directed, then M converges to sup M in the strong operator topology, i.e. for every $a \in E$ the net $\{T(a) : T \in M\}$ converges to $(\sup M)(a)$ in the norm topology of E.

13.3 A subspace $F \subset E$ is called an L-*ideal*, if it is the range of an L-projection. This is equivalent to the fact that F has a complement $F^\perp$ in E such that $\|m + n\| = \|m\| + \|n\|$ for all $m \in F$ and all $n \in F^\perp$. It follows that $F^\perp$ is uniquely determined and that $F^\perp$ is also an L-ideal. Moreover, if F and G are two L-ideals, then $F + G$ is closed and $F + G$ as well as $F \cap G$ are L-ideals.

13.4 The L-ideals in Banach spaces behave especially nice with respect to the extreme points of the unit ball of E. Let $B_1(E) = \{m \in E : \|m\| \le 1\}$ be the unit ball of E and let F be an L-ideal. Then we have

$$\operatorname{extr}(F \cap B_1(E)) = F \cap \operatorname{extr} B_1(E).$$

Moreover, if F and G are two L-ideals, then

$$(F + G) \cap B_1(E) = \operatorname{conv}((F \cap B_1(E)) \cup (G \cap B_1(E))).$$

13.5 Again, let $F \subset E$ be a closed subspace. If its polar $F^\circ \subset E'$ is an L-ideal in the topological dual E' of E equipped with its canonical norm, then F is called an M-*ideal* of E.
We record a few properties of M-ideals:

(i) Finite intersections and arbitrary closed linear spans of M-ideals are again M-ideals. Whence the M-ideals form a sublattice of the lattice of all closed subspaces of E. This lattice is complete, although arbitrary intersections in the lattice of M-ideals and the lattice of all closed subspaces do not agree.

By M(E) we denote the lattice of all M-ideals of E.

(ii) The sum of two M-ideals is closed.

(iii) A subspace $F \subset E$ is an M-ideal if and only if it has the following 3-ball property:
If $B(m_i, r_i) = \{m \in E : \|m - m_i\| < r_i\}$, i=1,2,3 are three open balls such that $B(m_1,r_1) \cap B(m_2,r_2) \cap B(m_3,r_3) \neq \emptyset$ and $B(m_i,r_i) \cap F \neq \emptyset$, then $B(m_1,r_1) \cap B(m_2,r_2) \cap B(m_3,r_3) \cap F \neq$ $\neq \emptyset$.

This 3-ball property may be used to show the following proposition (see also [GK 77] and [Be 79] in the case of function modules):

13.6 Proposition. *Let* $p : E \to X$ *be a bundle of Banach spaces with a compact base space* X *and let* $U \subset X$ *be an open subset. Then the subspace* $i(U) := \{\sigma \in \Gamma(p) : \sigma_{|X\setminus U} = 0\}$ *is an* M-*ideal of* $\Gamma(p)$.

Proof. We let $B_k := \{\sigma \in \Gamma(p) : \|\sigma_k - \rho_k\| < \varepsilon\}$ for $k = 1,2,3$ Suppose that $\sigma \in B_1 \cap B_2 \cap B_3$ and $\sigma_k \in B_k \cap i(U)$ for $k = 1,2,3$. Choose ε such that $0 < \varepsilon < \varepsilon - \|\rho_k - \sigma_k\|$ and $\varepsilon < \varepsilon_k - \|\rho_k - \sigma\|$ for $k = 1,2,3$ and let

$$V := \{x \in X : \|\sigma_k(x)\| < \varepsilon/2 \text{ for } k = 1,2,3\}.$$

Then V is open. Choose a continuous function $f : X \to [0,1]$ such that $f(x) = 0$ for $x \notin U$ and $f(x) = 1$ for $x \in X \setminus V$. Define a new section $\sigma' \in \Gamma(p)$ by $\sigma' := f\cdot\sigma$. We show that $\sigma' \in B_1 \cap B_2 \cap B_3 \cap i(U)$: Firstly, we have $\sigma' \in i(U)$ as $f(x) = 0$ for all $x \in X \setminus U$. In order to show that $\sigma' \in B_k$, consider $\|\sigma'(x) - \rho_k(x)\|$.

If $x \notin V$, then $\sigma'(x) = \sigma(x)$ and hence $\|\sigma'(x) - \rho_k(x)\| \leq \|\sigma - \rho_k\| \leq \varepsilon_k - \varepsilon$.

If $x \in V$, then we compute

$$\begin{aligned}\|\sigma'(x) - \rho_k(x)\| &= \|f(x)\cdot\sigma(x) - f(x)\cdot\rho_k(x) + \\ &\qquad + (f(x) - 1)\cdot(\rho_k(x) - \sigma_k(x) + \sigma_k(x)\| \\ &\leq |f(x)|\cdot\|\sigma(x) - \rho_k(x)\| + \\ &\qquad + |1 - f(x)|\cdot(\|\rho_k(x) - \sigma_k(x)\| + \|\sigma_k(x)\|) \\ &< f(x)\cdot(\varepsilon_k - \varepsilon) + (1 - f(x))\cdot(\varepsilon_k - \varepsilon + \varepsilon/2) \\ &\leq \varepsilon_k - \varepsilon/2.\end{aligned}$$

Thus we have shown that $\|\rho_k - \sigma'\| \leq \varepsilon - \varepsilon/2 < \varepsilon_k$,as desired. □

Hence, if $p : E \to X$ is a bundle of Banach spaces, we may "rediscover" the open subsets of X in $M(\Gamma(p))$ via the mapping i. The next result makes it even clearer what bundles should have to do with M-structure:

13.7 Let $F_1,\ldots,F_n$ be M-ideals of E such that $F_1 + \ldots + F_n = E$ and $F_1 \cap \ldots \cap F_n = \{0\}$. Then E is isometrically isomorphic to the cartesian product $\prod_{i=1}^{n} E/F_i$, equipped with the supremum norm.

We may interprete (13.7) as a representation of E by sections in a bundle $p : E \to \{1,\ldots,n\}$, where the stalk over i is just the quotient space E/F_i.

13.8 To generalize (13.7), we need again some notation . Let E be again be a Banach space and let F be an M-ideal of E. If there is an extreme point $p \in \text{extr } B_1(E')$ such that is F maximal among all M-ideals contained in ker(p) (or equivalently, if F^o is the smallest $\sigma(E',E)$-closed L-ideal containing p), then F is called *primitive*. Every primitive M-ideal F is a prime element in the lattice of all M-ideals, i.e. if G,H are two M-ideals, then $G \cap H \subset F$ implies $G \subset F$ or $H \subset F$. Moreover, every M-ideal G is the intersection of all primitive M-ideals containing G.

Especially, the lattice of all M-ideals is distributive (every lattice with the property that each element is a meet of prime elements is distributive.)

With Spec(E) we denote the set of all primitive M-ideals. If P is a primitive M-ideal, then we let $E_P := E/P$ and $a_P := a + P$ be the equivalence class of $a \in E$ modulo P. Note that E_P is a Banach space when equipped with the quotient norm and that we have a linear map

$$\begin{aligned} \hat{} : E &\to \Pi^\infty \{E_P : P \in \text{Spec}(E)\} \\ a &\to \hat{a} \quad \text{where } \hat{a}(P) := a_P. \end{aligned}$$

13.9 Proposition. The mapping $\hat{} : E \to \Pi^\infty E_P$ is an isometry.

Proof. Obviously, $\hat{}$ is a contraction. In order to show that $\|\hat{a}\| \geq \|a\|$, let $a \in E$. Then we can find an extreme point $p \in \text{extr } B_1(E')$ such that $\|a\| = p(a)$. Let P be the maximal M-ideal contained in ker(p). By duality, we may identify $(E/P)'$ with P^o. Then we obtain $\|a\| = p(a) = p(a_P) \leq \|p\| \cdot \|a_P\| = \|a_P\| \leq \|\hat{a}\|$. □

13.10 We now return to Banach bundles. Let $p : \mathcal{E} \to X$ be a bundle of Banach spaces and let $U \subset X$ be an open subset. By (13.6), the set $i(U) = N_{X\setminus U} = \{ \sigma \in \Gamma(p) : \sigma_{/X\setminus U} = 0\}$ is an M-ideal. As the primitive M-ideals are prime, we may equip Spec $\Gamma(p)$ with the hull-kernel topology. Now (12.16) applies to the lattice of all M-ideals :

13.11 Proposition. *Let* $p : E \to X$ *be a bundle of Banach spaces over a compact base space* X. *Then for every primitive* M-*ideal* P *there is an unique element* $\mathrm{fix}(P) \in X$ *such that* $N_{\mathrm{fix}(P)} \subset P$. *Moreover, the mapping* $\mathrm{fix} : \mathrm{Spec}\,\Gamma(p) \to X$ *is continuous, where* Spec $\Gamma(p)$ *carries the hull-kernel topology.* □

13.12 Let us consider again a locally C(X)-convex C(X)-module E, where E is a Banach space and where X is compact. As in section 7, we let $I_x = \{f \in C(X) : f(x) = 0\}$ and $N_x = \overline{I_x \cdot E}$. From (7.6) we conclude

$$(*) \qquad f\cdot a - f(x)\cdot a \in N_x \qquad \text{for all } x \in X,\ f \in C(X),\ a \in E.$$

Now we know from (7.19) that E is isometrically isomorphic to $\Gamma(p)$, where $p : E \to X$ is a bundle of Banach spaces. Under this isomorphism, N_x corresponds to $\{\sigma \in \Gamma(p) : \sigma(x) = 0\}$. Moreover, every continuous function $f \in C(X)$ yields a continuous function $\Phi(f) \in C_b(\mathrm{Spec}\,E)$ by $\Phi(f)(P) := f(\mathrm{fix}(P))$. As fix(P) is the unique element of X with $N_x \subset P$, the relation $(*)$ implies

$$(**) \qquad f\cdot a - \Phi(f)(P)\cdot a \in P \text{ for all } P \in \mathrm{Spec}\,E$$

or

$$({}^{**}_{*}) \qquad \Phi(f)\cdot a - \Phi(f)(P)\cdot a \in P \text{ for all } P \in \mathrm{Spec}\,E.$$

In $({}^{**}_{*})$ of course, we defined $\Phi(f)\cdot a := f\cdot a$.
The following important Dauns-Hofmann-Kaplansky multiplier theorem shows that we may define $f\cdot a$ for <u>every</u> $f \in C_b(\mathrm{Spec}\,E)$ and every $a \in E$ in such a manner that $({}^{**}_{*})$ remains valid. For a proof of this result we refer to [AE 72], [EO 74] or to [Be 79]:

13.13 Theorem. *Every Banach space* E *is a* C_b(Spec E)*-module such that for every* f $\in$ C_b(Spec E) *and every* P $\in$ Spec E *we have* $f \cdot a - f(P) \cdot a \in P$. □

13.14 Corollary. *Let* E *be a Banach space. Then*

(i) E *is a reduced* C_b(Spec E)*-locally convex* C_b(Spec E)*-module.*

(ii) If X *is any topological space, then* E *is a reduced* C(X)*-locally convex* C_b(X)*-module if and only if there is an isometric homomorphism of Banach algebras* $\Phi : C_b(X) \to C_b(\mathrm{Spec}\, E)$ *such that* $\Phi(f) \cdot a = f \cdot a$ *for all* $a \in E$.

Proof. Using (13.9) and the proof of (1.6.(x)) we see that E is C_b(Spec E)-locally convex. Moreover, the C_b(Spec E)-module E is reduced: Let $0 \neq f \in C_b(\mathrm{Spec}\, E)$. Then $f(P) \neq 0$ for some $P \in \mathrm{Spec}\, E$. Pick any $a \in E \setminus P$. Then $f(P) \cdot a \notin P$. As $f \cdot a - f(P) \cdot a \in P$ by (13.13), we conclude that $f \cdot a \neq 0$.

For a proof of (ii), assume that E is a reduced C(X)-locally convex C_b(X)-module. We may assume w.l.o.g. that X is compact. From (7.19) and (7.22) we conclude that there is a reduced bundle $p : \mathcal{E} \to X$ such that $E \simeq \Gamma(p)$.

We claim that for reduced bundles the mapping fix : Spec $\Gamma(p) \to X$ has dense image: Indeed, let $x \in X$ be an element such that $E_x = p^{-1}(x) \neq \{0\}$. Then $N_x = \{\sigma \in \Gamma(p) : \sigma(x) = 0\}$ is an M-ideal which is different from $\Gamma(p)$. By (13.8) we find a primitive M-ideal $P \in \mathrm{Spec}\, \Gamma(p)$ such that $N_x \subset P$. Clearly, fix(P) = x.

Now the mapping $\Phi : f \to f \circ \mathrm{fix} : C_b(X) \to C_b(\mathrm{Spec}\, \Gamma(p))$ sends $C_b(X)$ isometrically onto a closed subalgebra of $C_b(\mathrm{Spec}\, \Gamma(p)) = C_b(\mathrm{Spec}\, E)$. From (**) and (13.13) we obtain

$$f \cdot a - \Phi(f) \cdot a = f \cdot a - \Phi(f)(P) \cdot a + \Phi(f)(P) \cdot a - \Phi(f) \cdot a$$
$$\in P - P = P$$

for every $P \in \text{Spec } E$. As the intersection of all primitive M-ideals is 0, we obtain $f\cdot a = \Phi(f)\cdot a$. - The converse follows from (i). □

From (7.24) we know that we may identify $C_b(\text{Spec } E)$ with a closed subalgebra of $\mathcal{B}(E)$ via the mapping $f \to T_f$, $T_f(a) = f\cdot a$. This gives rise to the following definition:

13.15 Definition. Let E be a Banach space. The image of the mapping $f \to T_f : C_b(\text{Spec } E) \to \mathcal{B}(E)$ is called the *(topological) center of* E, denoted by $Z_t(E)$. □

There are various other characterizations of $Z_t(E)$ which we shall list. But firstly, we need a definition:

13.16 Definition. Let $T : E \to E$ be a bounded operator.
(i) T is called M-*bounded*, if there is a real number $r \in \mathbb{R}$ such that T(a) is contained in every ball of radius r which contains a.
(ii) T is called a *multiplier*, if there is a mapping

$$a_T : \text{extr } B_1(E') \to \mathbb{K}$$

such that for every $p \in \text{extr}.B_1(E')$ we have $T'(p) = a_T(p)\cdot p$, i.e. every extreme point of the dual unit ball is an eigenvector for the dual operator $T' : E' \to E'$.
(iii) If T and S are two multipliers on E, then S is called an *adjoint of* T, if $\overline{a_T} = a_S$, where $\bar{}$ denotes complex conjugation. □

From [AE 72] for the real case and [Be 79] for the arbitrary case we draw the following conclusion:

13.17 Theorem. *Let* $T : E \to E$ *be a bounded operator. Then*
(i) T *is* M-*bounded if and only if* T *is a multiplier.*

(ii) If the dual operator T' : E' → E' *belongs to the Cunningham algebra* Cu(E') *of* E', *then* T *is a multiplier.* □

13.18 Theorem. *Let* T : E → E *be a bounded operator. Then* T *belongs to the center* $Z_t(E)$ *if and only if* T *is a multiplier which admits an adjoint.*

If E *is a real Banach space, then the following conditions are equivalent:*

(i) T ∈ $Z_t(E)$.

(ii) T is M-bounded.

(iii) T is a multiplier.

(iv) T' ∈ Cu(E'). □

13.19 Conclusion. *Let* E *be a Banach space, let* X *be a compact topological space and let us assume that* E *is a* C(X)*-module.*

(i) E *is locally* C(X)*-convex if and only if the operators* a → f·a : E → E *belong to* $Z_t(E)$, f ∈ C(X).

(ii) E *is a reduced locally* C(X)*-convex* C(X)*-module if and only if* X *is a quotient of the Stone-Čech compactification* ß(Spec E) *of the space of all primitive* M*-ideals, equipped with the hull - kernel topology.* □

As a consequence, the space ß(Spec E) is the largest compact space which can serve as a base space of a bundle in which we can have a reduced representation of E by sections.

14. An adequate M-theory for Ω-spaces.

In this section we shall deal with a straightforward generalization of the ideas in section 13.

14.1 Definition. Let E be a topological Ω-Banach space. A closed subspace $M \subset E$ is called an *M-Ω-ideal,* if it is an M-ideal and an Ω-ideal at the same time. By $M_\Omega(E)$ we denote the set of all M-Ω-ideals. □

From (6.6) and (13.5.(i)) we obtain

14.2 Proposition. *Finite intersections and arbitrary closed linear spans of* M-Ω-*ideals leads again to* M-Ω-*ideals. Hence* $M_\Omega(E)$ *is a complete lattice, which is a sublattice of* $M(E)$ *and therefore distributive.* □

14.3 Proposition. *For every* M-*ideal (* Ω-*ideal)* $F \subset E$ *there is a largest* M-Ω-*ideal* $k_\Omega(F)$ *(*$k_M(F)$*) contained in* F. *Whence we have two kernel operators*

$$k_\Omega : M(E) \to M_\Omega(E)$$

and

$$k : Id_\Omega(E) \to M_\Omega(E) \qquad \square$$

Note that by (14.2) every closed subspace $F \subset E$ contains a largest M-Ω-ideal. This leads to

14.4 Definition. An M-Ω-ideal $P \subset E$ is called *primitive,* if there is an extreme point $p \in \operatorname{extr} B_1(E')$ of the dual unit ball such that P is the largest M-Ω-ideal of E contained in ker(p). By $\operatorname{Spec}_\Omega(E)$ we denote the collection of all primitive M-Ω-ideals. □

14.5 Proposition. *(i) The mapping* k_Ω *maps* Spec E *onto* $\operatorname{Spec}_\Omega$ E.

(ii) $\operatorname{Spec}_\Omega$ E *consists of prime elements of* $M_\Omega(E)$ *only.*

(iii) If we equip Spec E *and* $\operatorname{Spec}_\Omega$ E *with their hull-kernel topologies, then the restriction of* k_Ω *to* Spec E *is continuous.*

Proof. The first assertion is an obvious consequence of the definitions.

(ii) Let $P \in \operatorname{Spec}_\Omega E$ and pick any $Q \in \operatorname{Spec} E$ with $P = k_\Omega(Q) \subset Q$. If $M,N \in M_\Omega(E)$ are M-Ω-ideals with $M \cap N \subset P$, then we conclude $M \cap N \subset Q$, hence w.l.o.g. $M \subset Q$ as Q is prime. But this yields $k_\Omega(M) = M \subset k_\Omega(Q) = P$.

(iii) Let $A \subset \operatorname{Spec}_\Omega E$ be closed. Then there is an M-Ω-ideal M of E such that $A = \{P \in \operatorname{Spec}_\Omega E : M \subset P\}$. An easy calculation shows that $k_\Omega^{-1}(A) = \{P \in \operatorname{Spec} E : M \subset P\}$ and therefore $k_\Omega^{-1}(A)$ is closed in Spec E. This means that k_Ω is continuous. □

The following result is a consequence of the Dauns-Hofmann-Kaplansky multiplier theorem (13.13):

14.6 Theorem. *Let* E *be a topological* Ω*-Banach space. Then* E *is a* $C_b(\operatorname{Spec}_\Omega E)$*-*$\Omega$*-module which is locally* $C_b(\operatorname{Spec}_\Omega E)$*-convex and reduced. Moreover, for every* $P \in \operatorname{Spec}_\Omega E$*, every* $a \in E$ *and every* $f \in C_b(\operatorname{Spec}_\Omega E)$ *we have*

$$f\cdot a - f(P)\cdot a \in P$$

Proof. By (14.5.(iii)), the Banach algebra $C_b(\operatorname{Spec}_\Omega E)$ may be identified with a closed subalgebra of $C_b(\mathrm{Spec}\, E)$ via the mapping $f \to f \circ k_\Omega : C_b(\mathrm{Spec}_\Omega\, E) \to C_b(\mathrm{Spec}\, E)$. Define an action of $C_b(\mathrm{Spec}_\Omega\, E)$ on E by

$$f \cdot a := (f \circ k_\Omega) \cdot a.$$

By (13.14), the Banach space E becomes a reduced locally $C_b(\mathrm{Spec}_\Omega\, E)$--convex $C_b(\mathrm{Spec}_\Omega\, E)$-module in this way. Moreover, by (13.13) we have

$$f \cdot a - f(k_\Omega(Q)) \cdot a \in Q$$

for every $Q \in \mathrm{Spec}\, E$.

Now let $P \in \mathrm{Spec}_\Omega\, E$. From (13.8) we know that $P = \cap \{Q \in \mathrm{Spec}\, E : P \subset Q\}$. If $f \in C_b(\mathrm{Spec}_\Omega\, E)$ is given, then f is constant on the closure of $\{P\}$. As the closure of $\{P\}$ is the set $\{P' \in \mathrm{Spec}_\Omega\, E : P \subset P'\}$, we obtain $f(P) = f(P')$ whenever $P \subset P'$. Thus, if $Q \in \mathrm{Spec}\, E$ is given and if $P \subset Q$, then $P \subset k_\Omega(Q)$ and therefore $f(P) = f(k_\Omega(Q))$. This yields

$$f \cdot a - f(P) \cdot a \in Q$$

for all $Q \in \mathrm{Spec}\, E$ with $P \subset Q$ and thus

$$f \cdot a - f(P) \cdot a \in P = \cap \{Q \in \mathrm{Spec}\, E : P \subset Q\}.$$

It remains to show that E is a $C_b(\mathrm{Spec}_\Omega\, E)$-$\Omega$-module. So, let $f \in C_b(\mathrm{Spec}_\Omega\, E)$. Then for every $a \in E$ we have

$$\begin{aligned} a \in f^\perp \text{ iff } & f \cdot a = 0 \\ \text{iff } & f(P) \cdot a \in P \text{ for all } P \in \mathrm{Spec}_\Omega\, E \\ \text{iff } & a \in P \text{ for all } P \in f^{-1}(\mathbb{K} \setminus \{0\}) \\ \text{iff } & a \in \cap f^{-1}(\mathbb{K} \setminus \{0\}), \end{aligned}$$

i.e. $f^\perp = \cap \{P \in \mathrm{Spec}_\Omega\, E : f(P) \neq 0\}$. As an intersection of Ω-ideals yields again an Ω-ideal by (6.6), we conclude that $f^\perp$ is an Ω-ideal for every $f \in C_b(\mathrm{Spec}_\Omega\, E)$. □

Substituting $M(E)$ by $M_\Omega(E)$ and Spec E by Spec_Ω E, the same proofs as in (13.11) and (13.14) yield:

14.7 Proposition. *Let* $p : E \to X$ *be a bundle of* Ω*-Banach spaces over a compact base space* X. *Then for every* $P \in \mathrm{Spec}_\Omega\, \Gamma(p)$ *there is an unique element* $\mathrm{fix}(P) \in X$ *such that* $P \subset N_{\mathrm{fix}(p)}$. *Moreover, the mapping* $\mathrm{fix} : \mathrm{Spec}_\Omega\, \Gamma(p) \to X$ *is continuous.* □

14.8 Proposition. *If* X *is any compact topological space and if* E *is an* Ω*-Banach space, then* E *is a reduced and locally* $C_b(X)$*-convex* $C_b(X)$-Ω*-space if and only if there is an isometric homomorphism of Banach algebras* $\Phi : C_b(X) \to C_b(\mathrm{Spec}_\Omega\, E)$ *such that for all* $f \in C(X)$ *and all* $a \in E$ *we have* $\Phi(f)\cdot a = f\cdot a$. □

Again, (7.24) tells us that we may identify $C_b(\mathrm{Spec}_\Omega\, E)$ with a closed subalgebra of $B(E)$:

14.9 Definition. The image of the mapping $f \to T_f : C_b(\mathrm{Spec}_\Omega\, E) \to B(E)$ is called the *topological* Ω*-center of* E, denoted by $Z_{t,\Omega}(E)$. □

The following result is parallel to (13.19):

14.10 Theorem. *Let* E *be a topological* Ω*-Banach space, let* X *be a compact space and let us assume that* E *is a* $C(X)$*-module.*

(i) E *is a locally* $C(X)$*-convex* $C(X)$-Ω*-module if and only if the operators* $a \to f\cdot a : E \to E$ *belong to* $Z_{t,\Omega}(E)$ *for every* $f \in C(X)$.

(ii) E *is a reduced and locally* $C(X)$*-convex* $C(X)$-Ω*-module if and only if* X *is a quotient of the Stone-Čech compactification* $\beta(\mathrm{Spec}_\Omega\, E)$ *via a mapping* $\Phi : \beta(\mathrm{Spec}_\Omega\, E) \to X$ *such that* $f\cdot a - f(\Phi(P))\cdot a \in P$ *for all* $P \in \mathrm{Spec}_\Omega\, E$. □

Thus, as in (13.19), the space $\beta(\mathrm{Spec}_\Omega\, E)$ is the largest compact base space over which a representation of the Ω-space E by all sections in a bundle $p : E \to X$ is possible.

We conclude this sections with a few remarks concerning Banach lattices, Banach algebras and C^*-algebras. From (7.26) and (7.28) we obtain:

14.11 Proposition. *If* E *is a Banach algebra or a Banach lattice, then* $Z_{t,\Omega}(E) = Z_t(E) \cap Z_\Omega(E)$. □

If E is even a C^*-algebra, then we conclude from [Be 79] and [AE 72] that the M-ideals of E are exactly the closed two-sided ideals of the algebra E. Moreover, an ideal is primitive in the sense of M-ideals if and only if it is primitive as an ideal of the C^*-algebra E. Hence, in this case we have

$$Z(E) := Z_t(E) = Z_\Omega(E) = Z_{t,\Omega}(E)$$

and

$$\mathrm{Spec}\, E = \mathrm{Spec}_\Omega\, E$$

If we apply (14.10) to this situation, we obtain:

14.12 Corollary. (Dauns - Hofmann) *Let* E *be a unital* C^**-algebra and let* X *be the Stone-Čech compactification of* Spec E, *equipped with the hull-kernel topology. Then there is a bundle* $p : E \to X$ *of* C^**-algebras such that* E *is isometrically isomorphic to the* C^**-algebra* $\Gamma(p)$ *of all continuous sections of* p. □

15. Duality

The material represented in the rest of this paper was developped in order to give an useful representation of the dual space $\Gamma(p)'$ of the space of all sections in a bundle. Although I did not succeed to my satisfaction, I believe that many of the results discovered in this untertaking are interesting in themselves.

An "optimal" representation of linear functionals on $\Gamma(p)$ would be the following: Given a continuous linear form $\phi : \Gamma(p) \to \mathbb{K}$, where $p : E \to X$ is a bundle with stalks $(E_x)_{x \in X}$, then find a family $(\phi_x)_{x \in X}$ of continuous linear functionals $\phi_x : E_x \to \mathbb{K}$ and a measure $\mu \in M(X)$ such that

$$\phi(\sigma) = \int_X \phi_x(\sigma(x))\, d\mu(x)$$

for all $\sigma \in \Gamma(p)$.

Of course, this requires that the mapping $x \to \phi_x(\sigma(x)) : X \to \mathbb{K}$ is μ-integrable for every $\sigma \in \Gamma(p)$. As this is always the case if the mapping $T(\sigma) : X \to \mathbb{K}$ defined by $T(\sigma)(x) := \phi_x(\sigma(x))$ is continuous and as an easy calculation shows that $T(f \cdot \sigma) = f \cdot T(\sigma)$ for all $f \in C(X)$ and all $\sigma \in \Gamma(p)$, we are led to a study of the space of all (continuous) $C(X)$-module homomorphisms $T : \Gamma(p) \to C(X)$, denoted by $\mathrm{Mod}(\Gamma(p))$. It turns out that there is a close relation between the "size" of $\mathrm{Mod}(\Gamma(p))$ and the topology on E. We shall find out that (with restrictions) the space $\mathrm{Mod}(\Gamma(p))$ is "big" if and only if E is Hausdorff and that $\mathrm{Mod}(\Gamma(p))$ is "very big" if and only if the mapping $x \to \|\sigma(x)\| : X \to \mathbb{R}$ is continuous for every $\sigma \in \Gamma(p)$.

15.1 We start with a list of notations which we shall use frequently.

Let $p : E \to X$ be a full bundle. For every subset $A \subset X$ we let $N_A = \{\sigma \in \Gamma(p) : \sigma/_A = 0\}$. Instead of $N_{\{x\}}$ we shall write N_x.

Recall from (2.7) that for completely regular base spaces X the quotient $\Gamma(p)/N_x$ may be identified with the stalk E_x and whence the dual space E_x' of E_x may be identified with the polar $N_x^o \subset \Gamma(p)'$ of N_x. We shall always make this identification. Hence equations like

$$\phi(\sigma) = \phi(\sigma(x))$$

will make sense, provided that $\phi \in E_x' = N_x^o$.

Similarly, if X is normal and if $A \subset X$ is compact, then $\Gamma_A(p)$ may be identified with $\Gamma(p)/N_A$, if $p : E \to X$ is a bundle of Banach spaces. In this case we have $N_A^o = \Gamma_A(p)'$.

Let $p : E \to X$ be a bundle of Banach spaces with a completely regular base space X. For every $x \in X$ let $B_x \subset \Gamma(p)'$ be the dual unit ball of the stalk E_x. If $A \subset X$ is a subset, we define

$$B_A = \bigcup_{x \in A} B_x.$$

Note that B_A is not the dual unit ball of $\Gamma_A(p)$ in general! Finally, we let

$$B_A^\times = B_A \setminus \{0\}.$$

15.2 Proposition. *Let $p : E \to X$ be a full bundle with a completely regular base space.*

(i) If $A \subset X$ is closed and if $x \in X \setminus A$, then $N_x + N_A = \Gamma(p)$. If $p : E \to X$ is a bundle of Banach spaces, then

(ii) If $x,y \in X$ *are distinct, then* $B_x \cap B_y = N_x^O \cap N_y^O = \{0\}$

(iii) If $A \subset X$ *is closed, then* $B_A = B_X \cap N_A^O$.

Proof. (i) Let $f : X \to [0,1]$ be a continuous function such that $f(x) = 0$ and $f(A) = \{1\}$. If $\sigma \in \Gamma(p)$ is a section of the bundle p, then $f\cdot\sigma \in N_x$ and $(1 - f)\cdot\sigma \in N_A$ and thus $\sigma = f\cdot\sigma + (1 - f)\cdot\sigma$ belongs to $N_x + N_A$.

(ii) follows immediatly from (i) by taking polars.

(iii) If x is an element of $A \subset X$, then N_x contains N_A. Thus, using the definitions, we obtain $B_x \subset B_X \cap N_x^O \subset B_X \cap N_A^O$, i.e. $B_A \subset B_X \cap N_A^O$. Conversely, assume that $0 \neq \phi \in B_X \cap N_A^O$. Then we can find an $x \in X$ such that $\phi \in B_x \cap N_A^O$. We have to show that x belongs to A. Assume, if possible, that x does not belong to A. Then (i) yields the contradiction $\phi \in B_x \cap N_A^O \subset N_x^O \cap N_A^O = (N_x + N_A)^O = \Gamma(p)^O = \{0\}$. □

From now on, we shall always equip B_X with the weak-$*$-topology induced by $\Gamma(p)'$.

15.3 Proposition. *If* X *is completely regular and if* $A \subset X$ *is compact, then* B_A *is compact.*

Proof. Let $(\phi_i)_{i\in I}$ be a convergent net contained in B_A and let $\phi = \lim_{i\in I} \phi_i$. We have to show that $\phi \in B_A$. Firstly, for every $i \in I$ there is an $x_i \in A$ such that $\phi_i \in B_{x_i}$. By the compactness of A there is a convergent subnet $(x_j)_{j\in J}$ of $(x_i)_{i\in I}$; let $x := \lim_{j\in J} x_j$. We show that $\phi \in B_x$.

Obviously, we have $\|\phi\| \leq 1$. Moreover, note that $\phi = \lim_{j\in J} \phi_j$.

Now let $\sigma \in N_x$. We have to show that $\phi(\sigma) = 0$. Thus, let $\varepsilon > 0$. Then there is a neighborhood V of x such that $\|\sigma(y)\| < \varepsilon$ for all $y \in V$. Whence we have eventually $\|\sigma(x_j)\| < \varepsilon$. This yields

$$
\begin{aligned}
|\phi(\sigma)| &= \lim_{j\in J} |\phi_j(\sigma)| \\
&= \lim_{j\in J} |\phi_j(\sigma(x_j))| \\
&\le \overline{\lim_{j\in J}} \, \|\phi_j\| \cdot \|\sigma(x_j)\| \\
&\le \overline{\lim_{j\in J}} \, \|\sigma(x_j)\| \\
&\le \varepsilon
\end{aligned}
$$

As $\varepsilon > 0$ was arbitrary, we obtain $\phi(\sigma) = 0$. □

By (15.2) we have a mapping $\gamma : \mathcal{B}_X^\times \to X$ defined by $\gamma(\phi) = x$ iff $\phi \in \mathcal{B}_x$, provided that X is completely regular. Since for every subset $A \subset X$ we have $\gamma^{-1}(A) = \mathcal{B}_A^\times$, (15.3) allows us to conclude:

15.4 Proposition. *Let* $p : E \to X$ *be a bundle with a compact base space* X. *Then* $\mathcal{B}_X$ *is compact. Moreover, the mapping* $\gamma : \mathcal{B}_X^\times \to X$ *is continuous.* □

Recall from section 12 that Cl(X) denotes the complete lattice of all closed subsets of X. If $p : E \to X$ is a bundle with a compact base space, then we have a mapping

$$
\begin{aligned}
\gamma^* : \mathrm{Cl}(X) &\to \mathrm{Cl}(\mathcal{B}_X) \\
A &\to \mathcal{B}_A
\end{aligned}
$$

Note that $\gamma^*(A) = \gamma^{-1}(A) \cup \{0\}$. This yields

15.5 Proposition. *If* $p : E \to X$ *is a bundle of Banach spaces with a compact base space* X, *then the mapping* $\gamma^* : \mathrm{Cl}(X) \to \mathrm{Cl}(\mathcal{B}_X)$ *preserves arbitrary intersections and finite unions. Especially,* γ^* *is continuous for the Scott-topologies on* Cl(X) *and* $\mathrm{Cl}(\mathcal{B}_X)$, *resp.* □

15.6 Proposition. *Let $p : E \to X$ be a bundle of Banach spaces with a compact base space X. If $U \subset B_X$ is open and if $A \subset B_X$ is closed, then the set $\{x \in X : A \cap B_x \subset U\}$ is open in X.*

Proof. Let $\mathcal{U} = \{C \in Cl(B_X) : A \cap C \subset U\}$. If $C \in \mathcal{U}$ and if $C' \subset C$ is a closed subset of C, then C' belongs also to $\mathcal{U}$. Moreover, let $\mathcal{D} \subset Cl(B_X)$ be down-directed (i.e. $C_1, C_2 \in \mathcal{D}$ implies the existence of $C_3 \in \mathcal{D}$ such that $C_3 \subset C_1 \cap C_2$) and assume that $\cap \mathcal{D} \in \mathcal{U}$. By the definition of $\mathcal{U}$ this means $\cap \mathcal{D} \cap A \subset U$. Hence the compactness of B_X allows us to find a $C \in \mathcal{D}$ such that $C \cap A \subset U$, i.e. $C \in \mathcal{U}$. Thus, $\mathcal{U}$ is open in the Scott-topology of $Cl(B_X)$. As γ^* is Scott-continuous, $(\gamma^*)^{-1}(\mathcal{U}) = \{C \in Cl(X) : A \cap B_C \subset U\}$ is open in the Scott-topology of Cl(X). Now recall from [Comp 80] that the mapping $x \to \{x\} : X \to Cl(X)$ is even continuous for the Lawson-topology on Cl(X). This implies that the set $\{x \in X : A \cap B_x \subset U\}$ is open in X. □

In the following, let B_1^o be the unit ball of $\Gamma(p)'$ and let Conv B_1^o be the lattice of all closed convex subsets of B_1^o. Recall that B_1^o is a continuous lattice when ordered by dual inclusion. In the next proposition however, the lattice theoretical operations refer to the normal set theoretical inclusion.

15.7 Proposition. *Let $p : E \to X$ be a bundle of Banach spaces with a compact base space X. Then*

(i) $\overline{\text{conv}}\ B_A = N_A^o \cap B_1^o$ *for all closed subsets* $A \subset X$.

(ii) The mapping $A \to N_A^o \cap B_1^o : Cl(X) \to \text{Conv } B_1^o$ *preserves arbitrary intersections and finite suprema. Especially, this mapping is Scott-continuous.*

Proof. (i) Let B_1 be the unit ball of $\Gamma(p)$. Recall that the mapping

$$\varepsilon_A : \Gamma(p) \to \Gamma_A(p)$$
$$\sigma \to \sigma/A$$

is a quotient map with kernel N_A. If B_1^A is the unit ball of $\Gamma_A(p)$, then $\varepsilon_A^{-1}(B_1^A) = \overline{N_A + B_1}$. Moreover, we have $\|\varepsilon_A(\sigma)\| \leq 1$ if and only if $\|\sigma(x)\| \leq 1$ for all $x \in A$. This implies that $\overline{N_A + B_1} =$ $= \{\sigma \in \Gamma(p) : \|\sigma(x)\| \leq 1$ for all $x \in A\}$. As $\mathcal{B}_A$ contains 0, we obtain

$$\begin{aligned}
\overline{\text{conv}}\, \mathcal{B}_A &= \mathcal{B}_A^{oo} \\
&= (\bigcup_{x\in X} \mathcal{B}_x)^{oo} \\
&= (\bigcap_{x\in X} \mathcal{B}_x^{o})^{o} \\
&= \{\sigma \in \Gamma(p) : \|\sigma(x)\| \leq 1 \text{ for all } a \in A\}^{o} \\
&= \overline{(N_A + B_1)}^{o} \\
&= (N_A + B_1)^{o} \\
&= N_A^{o} \cap B_1^{o}.
\end{aligned}$$

(ii) If $A, A' \subset X$ are closed, then

$$\begin{aligned}
N_{A\cup A'}^{o} \cap B_1^{o} &= \overline{\text{conv}}\, \mathcal{B}_{A\cup A'} \\
&= \overline{\text{conv}}\, (\mathcal{B}_A \cup \mathcal{B}_{A'}) \\
&= \text{conv}\, (\overline{\text{conv}}\, \mathcal{B}_A \cup \overline{\text{conv}}\, \mathcal{B}_{A'}) \\
&= \text{conv}((N_A^{o} \cap B_1^{o}) \cup (N_{A'}^{o} \cap B_1^{o})) \\
&= (N_A^{o} \cap B_1^{o}) \vee (N_{A'}^{o} \cap B_1^{o}) ,
\end{aligned}$$

Hence finite suprema are preserved.

Moreover, N_A is an M-ideal of $\Gamma(p)$ and therefore N_A^{o} is an L-ideal of $\Gamma(p)'$. Thus, from (13.4) we conclude that

$$\begin{aligned}
\text{extr}\, (N_A^{o} \cap N_{A'}^{o} \cap B_1^{o}) &= N_A^{o} \cap N_{A'}^{o} \cap \text{extr}\, B_1^{o} \\
&= \text{extr}\, (N_A^{o} \cap B_1^{o}) \cap \text{extr}\, (N_{A'}^{o} \cap B_1^{o}).
\end{aligned}$$

Using the Krein-Milman theorem, we obtain from (i) firstly the inclusion

$$\text{extr.}(N^o_A \cap N^o_{A'} \cap B^o_1) \subset \mathcal{B}_A \cap \mathcal{B}_{A'}$$

$$= \mathcal{B}_{A\cap A'} \qquad \text{(as } \mathcal{B}_{A\cap A'} \text{ is closed)}$$

and then

$$N^o_A \cap N^o_{A'} \cap B^o_1 \subset \overline{\text{conv}}(\mathcal{B}_{A\cap A'})$$
$$= N^o_{A\cap A'} \cap B^o_1 .$$

Conversely, $A \cap A' \subset A, A'$ implies $N_A, N_{A'} \subset N_{A\cap A'}$ and therefore $N^o_{A\cap A'} \subset N^o_A \cap N^o_{A'}$. This shows that finite intersections are preserved.

Finally, let $(A_i)_{i\in I}$ be a down-directed family of closed subsets of X. From (12.12) we conclude that $N_{\cap A_i} = (\cup N_{A_i})^-$. Taking polars, we obtain $\cap (N^o_{A_i} \cap B^o_1) = N^o_{\cap A_i} \cap B^o_1$. □

15.8 Corollary. *If* $p : E \to X$ *is a bundle with a compact base space* X *and if* A *and* B *are closed subsets of* X, *then* $N_A + N_B = N_{A\cap B}$.

Proof. Obviously, we have $N_A + N_B \subset N_{A\cap B}$.
Conversely, note that N_A and N_B are M-ideals. Therefore the sum $N_A + N_B$ is a closed subspace of $\Gamma(p)$ by (13.5(ii)). It remains to show that $N_A + N_B$ is dense in $N_{A\cap B}$. But this follows immediatly from (15.7(ii)). □

These results may be interpreted that the semicontinuity of the norm in a bundle $p : E \to X$ is somehow reflected in the semicontinuity of the mapping $A \to N^o_A \cap B^o_1 : \text{Cl}(X) \to \text{Conv } B^o_1$ or, if we wish, in the semicontinuity of the mapping $x \to \mathcal{B}_x : X \to \text{Cl}(\mathcal{B}_X)$. Therefore, we might expect that the points of continuity of the mapping norm $: E \to \mathbb{R}$ are "rediscovered" in points of continuity of these maps:

15.9 Proposition. *Once again, let* $p : E \to X$ *be a bundle of Banach spaces with a compact base space* X *and let* $x_o \in X$ *be a point. The following conditions are equivalent:*

(i) The mapping $x \to B_x : X \to Cl(B_X)$ is continuous at x_o for the Lawson-topology on $Cl(B_X)$.

(ii) If $W \subset B_X$ is open and if $B_{x_o} \cap W \neq \emptyset$, then the set $\{x \in X : B_x \cap W \neq \emptyset\}$ is a neighborhood of x_o.

(iii) The mapping $x \to \|\sigma(x)\| : X \to \mathbb{R}$ is continuous at x_o for every $\sigma \in \Gamma(p)$.

(iii') The mapping norm $: E \to \mathbb{R} : \alpha \to \|\alpha\|$ is continuous at every α with $p(\alpha) = x_o$.

(iv) If $M \subset X$ is a subset of X and if x_o belongs to the closure of M, then $B_{x_o} \subset \overline{B_M}$.

Proof. (i) $\to$ (ii) : If $W \subset B_X$ is open, then $\{A \in Cl(B_X) : A \cap W \neq \emptyset\} = \{A \in Cl(B_X) : A \not\subset B_X \setminus W\}$ is open in the Lawson-topology. Thus (i) implies (ii).

(ii) $\to$ (iii) : Let $\sigma \in \Gamma(p)$, let $\varepsilon > 0$ and let $U = \{x \in X : \|\sigma(x)\| > \|\sigma(x_o)\| - \varepsilon\}$. We have to show that U is a neighborhood of x_o.

Let $W = B_X \cap \{\phi \in \Gamma(p)' : |\phi(\sigma)| > \|\sigma(x_o)\| - \varepsilon\}$. Then W is open in B_X. Moreover, we may find an ϕ in the dual unit ball B_{x_o} of E_{x_o} such that $\phi(\sigma(x_o)) = \phi(\sigma) > \|\sigma(x_o)\| - \varepsilon$. Hence the set $W \cap B_{x_o}$ is not empty and by (ii) the set $V = \{x \in X : B_x \cap W \neq \emptyset\}$ is a neighborhood of x_o. We complete the proof of (iii) by showing that $V \subset U$:

Indeed, if $x \in V$, then $B_x \cap W \neq \emptyset$. Hence there is an $\phi \in B_x$ such that $|\phi(\sigma)| > \|\sigma(x_o)\| - \varepsilon$. As $\|\phi\| \leq 1$, we may conclude that $\|\sigma(x_o)\| - \varepsilon < |\phi(\sigma)| = |\phi(\sigma(x))| \leq \|\phi\| \cdot \|\sigma(x)\| \leq \|\sigma(x)\|$, i.e.

$x \in U$.

(iii) $\rightarrow$ (iv) : Assume that $x_o \in \overline{M}$ but $\mathcal{B}_{x_o} \not\subset \overline{\mathcal{B}_M}$. Let $V = \mathcal{B}_X \setminus \overline{\mathcal{B}_M}$. Then V is open in $\mathcal{B}_X$ and $V \cap \mathcal{B}_{x_o} \neq \emptyset$. Moreover, we have

$$(1) \qquad r\cdot\phi \in V \quad \text{for all } \phi \in V \text{ and all } 1 \leq |r| \leq \frac{1}{\|\phi\|}.$$

Indeed, as $\|r\cdot\phi\| \leq 1$, we have $r\cdot\phi \in \mathcal{B}_X$. Now assume that $r\cdot\phi \in \overline{\mathcal{B}_M}$. Then $|\frac{1}{r}| \leq 1$ implies $\phi \in r^{-1}\cdot\overline{\mathcal{B}_M} = (r^{-1}\cdot\mathcal{B}_M)^- \subset \overline{\mathcal{B}_M}$.

Now pick any $\phi \in V \cap \mathcal{B}_{x_o}$ and let $A = \phi^{-1}(1)$. Then A is a closed hyperplane of $\Gamma(p)$. Further, we have

$$(2) \qquad \bigcap_{\substack{\sigma\in A \\ \varepsilon>0}} \{\psi \in \mathcal{B}_X : |\psi(\sigma)| \geq 1 - \varepsilon\} \subset V$$

To show this inclusion, let ψ belong the the left hand side. Then $|\psi(\sigma)| \geq 1$ for all $\sigma \in A$ and whence $\ker \psi \cap A = \emptyset$. This means that $\ker \psi$ and $A = \phi^{-1}(1)$ are parallel hyperplanes and thus $\ker \psi = \ker \phi$. Therefore, we can find an element $r \in \mathbb{K}$ such that $\psi = r\cdot\phi$. Pick any $\sigma \in A$. Then $\psi(\sigma) = r\cdot\phi(\sigma) = r$, whence $|r| \geq 1$. Moreover, we have $\|\psi\| = |r|\cdot\|\phi\| \leq 1$, i.e. $|r| \leq \frac{1}{\|\phi\|}$. This implies $r\cdot\phi = \psi \in V$, as we started with an $\phi \in V$.
Note that all the sets $\{\psi \in \mathcal{B}_X : |\psi(\sigma)| \geq 1 - \varepsilon\}$ are closed. An easy compactness argument shows:

(3) There are sections $\sigma_1,\ldots,\sigma_n \in \Gamma(p)$ and $\varepsilon > 0$ with $\phi(\sigma_i) = 1$ and $\bigcap_{i=1}^{n} \{\psi \in \mathcal{B}_X : |\psi(\sigma_i)| > 1 - \varepsilon\} \subset V$.

Let C be the convex hull of $\sigma_1,\ldots,\sigma_n$. Then

$$\phi(\sigma) = 1 \text{ for all } \sigma \in C.$$

Moreover, as C is compact, we can find elements $\rho_1,\ldots,\rho_m \in C$ such that for every $\sigma \in C$ there is an $j \in \{1,\ldots,m\}$ with $\|\sigma - \rho_j\| < \varepsilon/3$.

Now

$$\begin{aligned} 1 - \varepsilon/3 &< 1 \\ &= \phi(\rho_j) \\ &= \phi(\rho_j(x_o)) \\ &\leq \|\phi\| \cdot \|\rho_j(x_o)\| \\ &\leq \|\rho_j(x_o)\| \end{aligned}$$

and the assumption (iii) imply that there is an open neighborhood U of x_o such that $1 - \varepsilon/3 < \|\rho_j(x)\|$ for all $x \in U$. If $\sigma \in C$, then $\|\sigma - \rho_j\| < \varepsilon/3$ for a certain j. Therefore for all $x \in U$ we have

$$\begin{aligned} \|\rho_j(x)\| &\leq \|\sigma(x) - \rho_j(x)\| + \|\sigma(x)\| \\ &< \varepsilon/3 + \|\sigma(x)\| \end{aligned}$$

i.e.

$$1 - \varepsilon/3 < \varepsilon/3 + \|\sigma(x)\|$$

As x_o belongs to the closure of M, we can pick an element $x_1 \in M \cap U$. Thus we have shown:

(4) There is an $x_1 \in M$ such that $1 - \frac{2}{3}\cdot\varepsilon < \|\sigma(x_1)\|$ for all $\sigma \in C$.

From now on, we work entirely in the stalk E_{x_1}. Let $B \subset E_{x_1}$ be the ball of radius $1 - \frac{2}{3}\cdot\varepsilon$ and with center 0 and let $C_{x_1} := \{\sigma(x_1) : \sigma \in C\}$. Then it is clear that $B \cap C_{x_1} = \emptyset$ and hence $0 \notin C_{x_1} + B$. By the Hahn - Banach theorem we can find a continuous linear functional $\psi : E_{x_1} \to \mathbb{K}$ such that $\|\psi\| = 1$ and $\ker \psi \cap (C_{x_1} + B) = \emptyset$.

Suppose that $\|\psi(\sigma)\| \leq 1 - \varepsilon$ for a certain $\sigma \in C$. As $\|\psi\| = 1$, we can find an element $\alpha \in B$ such that $|\psi(\alpha)| > 1 - \varepsilon$. If we multiply α with an appropriate $r \in \mathbb{K}$ with $|r| \leq 1$ we obtain the existence of $\alpha' \in B$ such that $\psi(\alpha') = \psi(\sigma)$. Hence ψ maps $\sigma(x_1) - \alpha' \in C_{x_1} + B$ onto 0, a contradiction. Thus, we may conclude:

(5) There is an element $\psi \in B_{x_1}$ such that $|\psi(\sigma)| > 1 - \varepsilon$ for all $\sigma \in C$.

As we have $|\psi(\sigma_i)| > 1 - \varepsilon$ for all $i \in \{1,\dots,n\}$, this ψ belongs to V. On the other hand, we have $\psi \in B_{x_1} \subset B_M$, contradicting the fact that $V \cap B_M = (B_X \setminus \overline{B_M}) \cap B_M = \emptyset$.

(iv) → (i): Let $\mathcal{U}$ be an ultrafilter on X converging to x_o. We have to show that $\lim_{\mathcal{U}} B_x = B_{x_o}$.
Firstly, note that

$$\{x_o\} = \bigcap_{M \in \mathcal{U}} \overline{M}$$

and

$$\lim_{\mathcal{U}} B_x = \bigcap_{M \in \mathcal{U}} \Big(\bigcup_{x \in M} B_x \Big)^{-} \qquad \text{(see (8.8))}$$

i.e.

$$\lim_{\mathcal{U}} B_x = \bigcap_{M \in \mathcal{U}} \overline{B_M}\ .$$

As (iv) implies $B_{x_o} \subset \lim_{\mathcal{U}} B_x$, it remains to check the other inclusion. Let A be any closed neighborhood of x_o. Then $A \in \mathcal{U}$ and therefore

$$\begin{aligned} \bigcap_{M \in \mathcal{U}} \overline{B_M} &\subset \bigcap \{\overline{B_A} : A \text{ is a closed neighborhood of } x_o\} \\ &= \bigcap \{B_A : A \text{ is a closed neighborhood of } x_o\} \\ &= B_{x_o} \qquad \text{by (11.5).} \end{aligned}$$

(iii') → (iii) is trivial.

(iii) → (iii'): Let $\alpha \in E$ belong to the stalk E_{x_o} over x_o and choose any section $\sigma \in \Gamma(p)$ with $\sigma(x_o) = \alpha$. Further, let $\varepsilon > 0$. We have to find an open neighborhood U of α such that

$$|\,\|\beta\| - \|\alpha\|\,| < \varepsilon \text{ for all } \beta \in U.$$

An easy application of the triangle inequality shows that $U := \{\beta : \|\beta - \sigma(p(\beta))\| < \varepsilon/2 \text{ and } p(\beta) \in V\}$ has the required property, where V is any open set around x_o such that $| \|\sigma(x)\| - \|\alpha\| | < \varepsilon/2$ for all $x \in V$. □

15.10 Definition. We say that a bundle of Banach spaces $p : E \to X$ has *continuous norm*, if the mapping norm : $\alpha \to \|\alpha\| : E \to \mathbb{R}$ is continuous. □

In the following proposition we show that the continuity of the norm may be expressed by the continuity of various other maps:

15.11 Theorem. *Let* $p : E \to X$ *be a bundle of Banach spaces over a compact base space* X. *Then the following statements are equivalent:*

(i) E has continuous norm.

(ii) The mapping $x \to \|\sigma(x)\| : X \to \mathbb{R}$ is continuous for every $\sigma \in \Gamma(p)$.

(iii) If $W \subset B_X$ is open, then the set $\{x \in X : W \cap E_x' \neq \emptyset\}$ is open in X .

(iv) For every subset $M \subset X$ we have $\overline{B_M} = B_{\overline{M}}$.

(v) The mapping $x \to B_x : X \to Cl(B_X)$ is continuous for the Lawson topology.

(vi) The mapping $x \to B_x : X \to Cl(B_1^o)$ is continuous for the Lawson topology.

(vii) The mapping $x \to B_x : X \to \mathrm{Conv}\, B_1^o$ is continuous for the Lawson topology.

(viii) The mapping $A \to N_A^o \cap B_1^o : Cl(X) \to \mathrm{Conv}\, B_1^o$ is continuous for the Lawson topology.

(ix) The mapping $A \to B_A : Cl(X) \to Cl(B_X)$ is continuous for the

Lawson topology.

Proof. The equivalences of (i), (ii), (iii), (iv) and (v) follow immediately from (15.9)

(v) → (vi): As the embedding $A \to A : Cl(\mathcal{B}_X) \to Cl(B_1^o)$ preserves arbitrary infima and suprema, it is Lawson continuous. Hence (v) implies (vi).

(vi) → (vii) follows from (12.7).

(vii) → (vi): The embeddings $x \to \{x\} : X \to Cl(X)$ and $A \to A$: Conv B_1^o → $Cl(B_1^o)$ are continuous, hence (vii) implies (vi).

(vi) → (v): The image of the mapping $x \to \mathcal{B}_x$ is contained in $Cl(\mathcal{B}_X)$. As the embedding $A \to A : Cl(\mathcal{B}_X) \to Cl(B_1^o)$ is also a topological embedding, (v) follows.

(iv) → (viii): We show that the mapping $A \to N_A^o \cap B_1^o$ preserves arbitrary suprema: Indeed, (15.7(i)) yields $N^o_{\overline{\cup A_i}} \cap B_1^o = \overline{\mathrm{conv}}(\mathcal{B}_{\overline{\cup A_i}}) = \overline{\mathrm{conv}}(\overline{\mathcal{B}_{\cup A_i}})$ (by (iv)) $= \overline{\mathrm{conv}}(\cup \mathcal{B}_{A_i}) = \overline{\mathrm{conv}}(\cup \overline{\mathrm{conv}}\, \mathcal{B}_{A_i}) =$ $= \overline{\mathrm{conv}}(\cup (N^o_{A_i} \cap B^o_1))$, i.e. $N^o_{\sup A_i} \cap B^o_1 = \sup(N^o_{A_i} \cap B^o_1)$. As this mapping always preserves arbitrary infima, it is Lawson--continuous by (12.4(ii)).

For (viii) → (ix) and (ix) → (v) use the arguments given in (vii) → (vi) → (v). □

We now develop a duality between "stalkwise convex" subsets of E and "stalkwise convex" subsets of $\mathcal{B}_X$. In the remainder of this section,

$p : E \to X$ <u>will always be a bundle of Banach spaces with a compact base space</u> X.

15.12 Lemma. *Let* $K \subset B_X$ *be closed and let* $\varepsilon > 0$. *Define*

$$K^{\varepsilon} := \{\alpha \in E : \operatorname{Re}\phi(\alpha) \leq \varepsilon \text{ for all } \phi \in K \cap B_{p(\alpha)}\}.$$

Then the restriction $p : K^{\varepsilon} \to X$ *of the projection* $p : E \to X$ *is still open.*

Proof. Let $\alpha \in K^{\varepsilon}$ and let U be an open neighborhood of α. We have to show that $p(U \cap K^{\varepsilon})$ is a neighborhood of $p(\alpha)$.
First of all, we may assume that U has the form

$$U = \{\beta \in E : \|\tau(p(\beta)) - \beta\| < r \text{ and } p(\beta) \in W\},$$

where $\tau \in \Gamma(p)$ is a section with $\tau(p(\alpha)) = \alpha$ and where W is an open neighborhood of $p(\alpha)$.
If $\tau = 0$, then the 0 of E_x is contained in $U \cap K^{\varepsilon}$ for every $x \in W$ and whence $p(U \cap K^{\varepsilon}) = W$
From now on, we assume that $\tau \neq 0$. By passing to a smaller r if necessary, we obtain $\|\tau\| - r > 0$. Choose a real number δ such that $\varepsilon < \delta < \varepsilon \cdot \frac{\|\tau\|}{\|\tau\| - r}$ and let $O = \{\phi \in B_X : \operatorname{Re}\phi(\alpha) < \delta\}$. Then O is open and so is the set $\{x \in X : K \cap B_x \subset O\}$ (see (15.6)). Since for all $\phi \in B_{p(\alpha)} \cap K$ the inequality $\operatorname{Re}\phi(\alpha) \leq \varepsilon < \delta$ holds, $p(\alpha)$ belongs to the open set $W \cap \{x \in X : K \cap B_x \subset O\}$. We claim that $W \cap \{x \in X : K \cap B_x \subset O\} \subset p(U \cap K^{\varepsilon})$:
Let $x_o \in W \cap \{x \in X : K \cap B_x \subset O\}$. Then the element $\frac{\varepsilon}{\delta} \cdot \tau(x_o)$ belongs in fact to K^{ε}: Indeed, let ϕ be any element of $K \cap B_{x_o}$. Then ϕ belongs to O, whence $\operatorname{Re}\phi(\alpha) = \operatorname{Re}\phi(\tau(x_o)) < \delta$, i.e. $\operatorname{Re}\phi(\frac{\varepsilon}{\delta} \cdot \tau(x_o)) < \varepsilon$. Moreover, we have $\|\frac{\varepsilon}{\delta} \cdot \tau(x_o) - \tau(x_o)\| = (1 - \frac{\varepsilon}{\delta}) \cdot \|\tau(x_o)\| <$ $< (1 - \frac{\|\tau\| - r}{\|\tau\|}) \cdot \|\tau\| = r$, i.e. $\frac{\varepsilon}{\delta} \cdot \tau(x_o) \in U$. We finally conclude

$\frac{\varepsilon}{\delta}\cdot\tau(x_o) \in U \cap K^{\varepsilon}$ and therefore $x_o \in p(U \cap K^{\varepsilon})$. □

15.13 **Proposition.** *Let* $K \subset E$ *be a subset such that* $K \cap E_x$ *is closed, convex and non-empty for every* $x \in X$. *If the restriction* $p : K \to X$ *of the projection* $p : E \to X$ *is still open, then for every* $\alpha \in K$ *there is a section* $\sigma \in \Gamma(p)$ *with* $\sigma(p(\alpha)) = \alpha$ *and* $\sigma(x) \in K$ *for every* $x \in X$.

Proof. <u>Step 1</u> If $\varepsilon > 0$ and if $\alpha_o \in K$ are given, then there is a section $\sigma \in \Gamma(p)$ such that $\sigma(p(\alpha_o)) = \alpha_o$ and such that for every $x \in X$ there is an $\alpha \in K \cap E_x$ satisfying $||\sigma(x) - \alpha|| < \varepsilon$.
(Proof of step 1: Let $\alpha_o \in K$ and let $x \in X$ be arbitrary. Then we may find a section $\sigma_x \in \Gamma(p)$ such that $\sigma_x(p(\alpha_o)) = \sigma_o$ and $\sigma_x(x) \in K \cap E_x$. Let

$$U_x := \{\beta \in E: ||\sigma_x(p(\beta)) - \beta|| < \varepsilon\}.$$

Then, by assumption, $V_x := p(U_x \cap K)$ is an open neighborhood of x. As X is compact and as the V_x, $x \in X$, cover X, we can find finitely many elements $x_1,\ldots,x_n \in X$ such that $X = V_{x_1} \cup \ldots \cup V_{x_n}$. Let $(f_i)_{i=1}^{n}$ be a partition of unity subordinate to the open cover $V_{x_1},\ldots,V_{x_n}$. We define a section $\sigma \in \Gamma(p)$ by

$$\sigma := \sum_{i=1}^{n} f_i\cdot\sigma_{x_i}.$$

Then

$$\begin{aligned}\sigma(p(\alpha_o)) &= \sum_{i=1}^{n} f_i(p(\alpha_o))\cdot\sigma_{x_i}(p(\alpha_o))\\ &= \sum_{i=1}^{n} f_i(p(\alpha_o))\cdot\alpha_o\\ &= \alpha_o.\end{aligned}$$

Moreover, if $x \in X$ is given, let $M := \{i : f_i(x) \neq 0\}$. Then we have $x \in V_{x_i}$ for every $i \in M$, as f_i vanishes outside V_{x_i}. Hence for every $i \in M$ we can find a $\beta_i \in K \cap E_x$ such that $\|\sigma_{x_i}(x) - \beta_i\| < \varepsilon$. Define

$$\beta := \sum_{i \in M} f_i(x) \cdot \beta_i$$

As $K \cap E_x$ is convex and as $\sum_{i \in M} f_i(x) = 1$, the element β belongs to $K \cap E_x$. Finally, we have

$$\begin{aligned} \|\sigma(x) - \beta\| &= \|\sum_{i \in M} f_i(x) \cdot \sigma_{x_i}(x) - \beta\| \\ &\leq \sum_{i \in M} f_i(x) \cdot \|\sigma_{x_i}(x) - \beta_i\| \\ &< \sum_{i \in M} f_i(x) \cdot \varepsilon \\ &= \varepsilon \end{aligned}$$

<u>Step 2</u> Let $\varepsilon > 0$ and let $\alpha_o \in K$. Assume that there is a section $\tau \in \Gamma(p)$ with $\tau(p(\alpha_o)) = \alpha_o$ and assume that for every $x \in X$ there is an $\alpha \in K \cap E_x$ such that $\|\tau(x) - \alpha\| < \varepsilon$. Then we can find a section $\tau' \in \Gamma(p)$ such that

(i) $\|\tau' - \tau\| \leq \varepsilon$.

(ii) $\tau'(p(\alpha_o)) = \alpha_o$.

(iii) For every $x \in X$ there is an $\alpha \in K \cap E_x$ satisfying $\|\alpha - \tau'(p(\alpha))\| < \varepsilon/2$.

(Proof of step 2: Let $x_o = p(\alpha_o)$ and let $x \in X$ be arbitrary. Then there is an $\alpha_x \in E_x \cap K$ with $\|\tau(x) - \alpha_x\| < \varepsilon$. We may assume that $\alpha_x = \alpha_o$ if $x = x_o$. Let τ_x be any continuous section such that $\tau_x(x) = \alpha_x$ and such that for every $y \in X$ there is an $\alpha' \in E_y \cap K$ satisfying $\|\tau_x(y) - \alpha'\| < \varepsilon/2$. As $\|\tau_x(x) - \tau(x)\| = \|\alpha_x - \tau(x)\| < \varepsilon$, we can find an open neighborhood U_x of x such that $\|\tau_x(y) - \tau(y)\| < \varepsilon$ for all $y \in U_x$. We may assume that $x_o \notin U_x$ if

$x \neq x_o$. Let $(f_x)_{x \in X}$ be a partition of unity subordinate to the open cover $(U_x)_{x \in X}$ of X. Then $f_{x_o}(x_o) = 1$ and $f_x(x_o) = 0$ for $x \neq x_o$. Define

$$\tau' := \sum_{x \in X} f_x \cdot \tau_x$$

Then $\tau'(x_o) = \alpha_o$ and $||\tau' - \tau|| \leq \varepsilon$ by some standard arguments we already used in the proof of the Stone-Weierstraß theorem (4.2). Moreover, as in the proof of step 1, we see that for every $x \in X$ there is an $\alpha \in K \cap E_x$ such that $||\tau'(x) - \alpha|| < \varepsilon/2$.)

<u>Step 3</u> For every $\alpha_o \in K$ there is a $\sigma \in \Gamma(p)$ with $\sigma(p(\alpha_o)) = \alpha_o$ and $\sigma(x) \in K$ for all $x \in X$.

(By induction, using step 1 and step 2, we can find a sequence $\tau_n \in \Gamma(p)$ such that

(i) $\tau_n(p(\alpha_o)) = \alpha_o$ for all $n \in \mathbb{N}$.

(ii) $||\tau_n - \tau_{n+1}|| \leq (\frac{1}{2})^n$ for all $n \in \mathbb{N}$.

(iii) For every $n \in \mathbb{N}$ and every $x \in X$ there is an $\alpha_{n,x} \in E_x \cap K$ such that $||\tau_n(x) - \alpha_{n,x}|| < (\frac{1}{2})^n$.

We compute that

$$\begin{aligned}||\alpha_{n,x} - \alpha_{n+1,x}|| &\leq ||\alpha_{n,x} - \tau_n(x)|| + ||\tau_n(x) - \tau_{n+1}(x)|| \\ &\quad + ||\tau_{n+1}(x) - \alpha_{n+1,x}|| \\ &\leq 3 \cdot (\tfrac{1}{2})^n.\end{aligned}$$

Hence $(\tau_n)_{n \in \mathbb{N}}$ and $(\alpha_{n,x})_{n \in \mathbb{N}}$ are Cauchy sequences. Let $\sigma := \lim_{n\to\infty} \tau_n$. Then for each $n \in \mathbb{N}$ we have $\tau_n(p(\alpha_o)) = \alpha_o$ and therefore $\sigma(p(\alpha_o)) = = \alpha_o$. Finally, if $x \in X$, then $\sigma(x) = \lim_{n\to\infty} \tau_n(x) = \lim_{n\to\infty} \alpha_{n,x} \in K \cap E_x$, as $K \cap E_x$ is closed.) □

15.14 Proposition. *Let* $K \subset B_X$ *be closed. Then for every* $x \in X$ *we have*

$$(K \cap B_x)^o = \{\sigma \in \Gamma(p) : \sigma(x) = \tau(x) \text{ for some } \tau \in K^o\}$$
$$= K^o + N_x$$
$$= \overline{\text{conv}}\,(K^o \cup N_x)$$

Especially, $K^o + N_x$ *is closed for every* $x \in X$

Proof. Firstly, we have

$$\{\sigma \in \Gamma(p) : \sigma(x) = \tau(x) \text{ for some } \tau \in K^o\} = K^o + N_x$$
$$\subset \overline{\text{conv}}\,(K^o \cup N_x)$$
$$\subset (K^o \cup N_x)^{oo}$$
$$= (K^{oo} \cap N_x^o)^o$$
$$= (K^{oo} \cap E_x')^o$$
$$= (K^{oo} \cap B_x)^o$$
$$\subset (K \cap B_x)^o.$$

Thus, it remains to show that $(K \cap B_x)^o \subset \{\sigma \in \Gamma(p) : \sigma(x) = \tau(x)$ for some $\tau \in K^o\}$.

Let $\sigma \in (K \cap B_x)^o$. Then

$$\sigma(x) \in \{\alpha \in E : \text{Re}\,\phi(\alpha) \le 1 \text{ for all } \phi \in K \cap B_{p(\alpha)}\}.$$

Hence, using (15.12) and (15.13), we can find a section $\tau \in \Gamma(p)$ such that

$$\sigma(x) = \tau(x)$$

and

$$\tau(y) \in \{\alpha \in E : \text{Re}\,\phi(\alpha) \le 1 \text{ for all } \phi \in K \cap B_{p(\alpha)}\}$$

for all $y \in X$.

Now let $\phi \in K$. Then ϕ belongs to $K \cap B_y$ for a certain $y \in X$ and whence $\text{Re}\,\phi(\tau) = \text{Re}\,\phi(\tau(y)) \le 1$. But this implies $\tau \in K^o$. □

15.15 Corollary. *If* $K \subset \mathcal{B}_X$ *is closed and if* $0 \in K$, *then*

(i) $(\overline{\text{conv}}\ K) \cap \mathcal{B}_x = \overline{\text{conv}}\ (K \cap \mathcal{B}_x)$ *for every* $x \in X$

(ii) $\bigcup_{x \in X} \overline{\text{conv}}\ (K \cap \mathcal{B}_x)$ *is closed.*

Proof. (i) From (15.14) we conclude that

$$\begin{aligned} \overline{\text{conv}}\ (K \cap \mathcal{B}_x) &= (K \cap \mathcal{B}_x)^{oo} \\ &= (\overline{\text{conv}}\ (K^o \cup N_x))^o \\ &= (K^o \cup N_x)^{ooo} \\ &= (K^o \cup N_x)^o \\ &= K^{oo} \cap N_x^o \\ &= \overline{\text{conv}}\ K \cap \mathcal{B}_x . \end{aligned}$$

(ii) follows from (i), as

$$\begin{aligned} \bigcup_{x \in X} \overline{\text{conv}}\ (K \cap \mathcal{B}_x) &= \bigcup_{x \in X} (\overline{\text{conv}}\ K) \cap \mathcal{B}_x \\ &= \mathcal{B}_X \cap \overline{\text{conv}}\ K. \quad \square \end{aligned}$$

We now go back to the discussion of subbundle as it was begun in section 8. We shall apply the results obtained in the present section in order to give a description of subbundles which uses duality.

Firstly, recall from (8.8) that a subbundle $F \subseteq E$ is completely determined by a "distribution" of closed subspaces $(F_x)_{x \in X}$ of the stalks such that the restriction of the projection $p: E \to X$ to $\bigcup_{x \in X} F_x$ is still open.

The next lemma is certainly well-known to everyone working in functional analysis:

15.16 Lemma. *Let* E *be a Banach space and let* $K \subset E'$ *be a subset such that*

(a) K is $\sigma(E',E)$-compact, convex and circled.

(b) $\|\phi\| \leq 1$ for all $\phi \in K$.

(c) If $0 \neq \phi \in K$, then $\phi/\|\phi\| \in K$.

If $a \in E$ *and* $\varepsilon > 0$ *are given such that* $|\phi(a)| < \varepsilon$ *for all* $\phi \in K$, *then there is an element* $b \in E$ *such that* $\|a - b\| < \varepsilon$ *and* $\phi(b) = 0$ *for all* $\phi \in K$.

Moreover, $\mathbb{K}\cdot K$ *is the* $\sigma(E',E)$*-closed subspace generated by* K.

Proof. Let $F \subset E'$ be the subspace generated by K. Then $F = \mathbb{K}\cdot K$, as K is convex and circled. Moreover, $F \cap \{\psi \in E' : \|\psi\| \leq 1\} = K$ by the assumptions (b) and (c). From the Krein-Smulian theorem and (a) we conclude that F is $\sigma(E',E)$-closed.

Thus, E/F^{o} is a Banach space and $(E/F^{o})'$ is isometrically isomorphic with $F^{oo} = F$. The dual unit ball of E/F^{o} may be identified with K. Now let $\pi : E \to E/F^{o}$ be the canonical projection. If $|\phi(a)| < \varepsilon$ for all $\phi \in K$, then $\|\pi(a)\| < \varepsilon$. As $\|\pi(a)\| = \inf\{\|a - b\| : b \in F^{o}\}$, there is a $b \in F^{O} = \{u \in E : \phi(u) = 0 \text{ for all } \phi \in K\}$ such that $||a - b|| < \varepsilon$. □

We now return to our bundle $p : E \to X$ of Banach spaces with a compact base space X.

15.17 Proposition. *Let* $K \subset B_X$ *be a closed set such that*

(a) $K_x := K \cap B_x$ is convex and circled for every $x \in X$

(b) If $0 \neq \phi \in K$, then $\phi / ||\phi|| \in K$.

Then $E_K = \{\alpha \in E : \phi(\alpha) = 0 \text{ for all } \phi \in K \cap B_{p(\alpha)}\}$ *is a subbundle of* E.

Proof. Obviously, $E_K \cap E_x = \{\alpha \in E_x : \phi(\alpha) = 0 \text{ for all } \phi \in K \cap B_x\}$ is a closed linear subspace of E_x. It remains to show that the restriction $p_{/E_K} : E_K \to X$ is open.
Thus, let $\alpha \in E_K$ and let $U \subset E$ be an open set around α. We have to show that $p(U \cap E_K)$ is a neighborhood of $p(\alpha)$. Firstly, we may assume without loss of generality that U has the form

$$U = \{\beta \in E : ||\sigma(p(\beta)) - \beta|| < \varepsilon \text{ and } p(\beta) \in W\},$$

where $\sigma \in \Gamma(p)$ is a section passing through α, where $\varepsilon > 0$ and where W is an open neighborhood of $p(\alpha)$. As in (15.12) we define

$$K^{\varepsilon/3} = \{\beta \in E : \operatorname{Re} \phi(\beta) \leq \varepsilon/3 \text{ for all } \phi \in K \cap B_{p(\beta)}\}.$$

As K is circled, we may write

$$K^{\varepsilon/3} = \{\beta \in E : |\phi(\beta)| \leq \varepsilon/3 \text{ for all } \phi \in K \cap B_{p(\beta)}\}.$$

As $\alpha \in K^{\varepsilon/3}$, using (15.12) and (15.13), we can find a section $\rho \in \Gamma(p)$ such that $\rho(p(\alpha)) = \alpha$ and $\rho(x) \in K^{\varepsilon/3}$ for all $x \in X$. As $\rho(p(\alpha)) - \sigma(p(\alpha)) = \alpha - \alpha = 0$ and as the mapping norm : $E \to \mathbb{R}$ is upper semicontinuous, the set

$$W' := \{x \in W : \|\rho(x) - \sigma(x)\| < \varepsilon/3\}$$

is an open neighborhood of $p(\alpha)$. We claim that $W' \subset p(U \cap E_K)$: Indeed, if $x \in W'$, then

$$|\phi(\rho(x))| \le \varepsilon/3 < \varepsilon/2 \text{ for all } \phi \in K \cap B_x$$

and

$$\|\sigma(x) - \rho(x)\| < \varepsilon/3$$

whence for $\phi \in K \cap B_x$ we have

$$\begin{aligned} |\phi(\sigma(x))| &\le |\phi(\sigma(x) - \rho(x))| + |\phi(\rho(x))| \\ &< \|\phi\| \cdot \|\sigma(x) - \rho(x)\| + \varepsilon/2 < \varepsilon \end{aligned}$$

Applying (15.16) to the Banach space E_x and the compact set $K \cap B_x$, we obtain an element $\alpha \in E_x$ such that $\|\alpha - \sigma(x)\| < \varepsilon$ and $\phi(\alpha) = 0$ for all $\phi \in K \cap B_x$. As $p(\alpha) = x$ and as $\alpha \in U \cap E_K$, the proof is complete. □

15.18 Theorem. *Let $p : E \to X$ be a bundle of Banach spaces with a compact base space X. Then the mapping*

$$F \to F^{\circ} \cap B_X =: K_F$$

$$(\text{resp. } F \to \bigcup_{x \in X} (F \cap E_x)^{\circ} \cap B_X =: K_F)$$

is a bijection between the set of all closed submodules of $\Gamma(p)$ (resp. subbundles of E) onto the set of all closed subsets $K \subset B_X$ such that

(a) $K \cap B_x$ is convex and circled for every $x \in X$.

(b) If $0 \neq \phi \in K$, then $\phi/\|\phi\| \in K$.

The inverse of this mapping is given by

$$K \to F_K := \{\sigma \in \Gamma(p) : \phi(\sigma(x)) = 0 \text{ for all } x \in X,\ \phi \in K \cap B_x\}$$

(resp. $K \to E_K := \{\alpha \in E : \phi(\alpha) = 0 \text{ for all } \phi \in K \cap B_{p(\alpha)}\}$ *).*

Moreover, the bijections given here and the bijections given in (8.6) commute when composed in the right order.

Proof. First of all, note that we may (and do) identify $(F \cap E_x)^o$ with $(\Gamma(p_{/F}) + N_x)^o = \Gamma(p_{/F})^o \cap N_x^o$. Hence we obtain

$$\begin{aligned} K_F &= \bigcup_{x \in X} (F \cap E_x)^o \cap B_X \\ &= \bigcup_{x \in X} (\Gamma(p_{/F})^o \cap N_x^o) \cap B_X \\ &= \Gamma(p_{/F})^o \cap B_X \\ &= K_{\Gamma(p_{/F})}. \end{aligned}$$

Obviously, we have $F_K = \Gamma(p_{/E_K})$. Therefore, it is enough to show that the mappings $F \to K_F$ and $K \to E_K$ are inverse to each other: Let $F \subset E$ be a subbundle. Then

$$\begin{aligned} E_{K_F} &= \{\alpha \in E : \phi(\alpha) = 0 \text{ for all } \phi \in K_F \cap B_{p(\alpha)}\} \\ &= \{\alpha \in E : \phi(\alpha) = 0 \text{ for all } \phi \in (F \cap E_{p(\alpha)})^o\} \\ &= \{\alpha \in E : \alpha \in (F \cap E_{p(\alpha)})^{oo}\} \\ &= F \end{aligned}$$

Conversely, if $K \subset B_X$ satisfies (a) and (b), then

$$\begin{aligned} K_{E_K} &= \bigcup_{x \in X} (E_K \cap E_x)^o \cap B_X \\ &= \bigcup_{x \in X} \{\alpha \in E_x : \phi(\alpha) = 0 \text{ for all } \phi \in K \cap B_x\}^o \cap B_X \\ &= B_X \cap \bigcup_{x \in X} \mathbb{K} \cdot (K \cap B_x) \\ &= K, \end{aligned}$$

where we used (15.16) to establish the equality $\mathbb{K} \cdot (K \cap B_x) = \{\alpha \in E_x : \phi(\alpha) = 0 \text{ for all } \phi \in K \cap B_x\}^o$. □

15.19 Corollary. *Let $p : E \to X$ be a bundle of Banach spaces with a compact base space X. If $(F_x)_{x \in X}$ is a family of closed linear subspaces of $(E_x)_{x \in X}$, then for every $x_o \in X$ and every $\alpha \in F_{x_o}$ there is a section $\sigma \in \Gamma(p)$ such that*

$$\sigma(x_o) = \alpha$$

and $$\sigma(x) \in F_x \quad \text{for all } x \in X$$

if and only if $\bigcup_{x \in X} F_x^o \cap B_X$ *is closed.* □

We conclude this section with a remark concerning M-ideals of $\Gamma(p)$. Firstly, let M be a primitive M-ideal. Then we recall from that there is an element $x \in X$ such that $N_x \subset M$. This implies that M is a submodule of $\Gamma(p)$: Indeed, let $f \in C(X)$ and let $\sigma \in M$. Then $f \cdot \sigma - f(x) \cdot \sigma \in N_x \subset M$ and $f(x) \cdot \sigma \in M$, hence $f \cdot \sigma$ belongs to M, too.

As every M-ideal is the intersection of primitive M-ideals, we obtain:

15.20 Proposition. *Every M-ideal $M \subset \Gamma(p)$ is a C(X)-submodule of $\Gamma(p)$.* □

Using the 3-ball property of M-ideals, we can show the following (see [Be 79, p.86]):

15.21 Theorem. *Let $p : E \to X$ be a bundle of Banach spaces over a compact base space X. A closed linear subspace $M \subset \Gamma(p)$ is an M-ideal of $\Gamma(p)$ if and only if there is a subbundle $E_M \subset E$ such that*

(i) $M = \Gamma(p_{/E_M})$

(ii) E_M *is "stalkwise" an M-ideal of E, i.e. the linear subspace $E_M \cap E_x$ is an M-ideal of E_x.* □

16. The closure of the "unit ball" of a bundle and separation axioms

As we already noticed in example (5.16), not all bundle spaces have to be Hausdorff. The same example shows that the closure of the 0-section may contain a whole line in some stalk and the same is true for the closure of the "unit ball" $\{\alpha \in E : \|\alpha\| \leq 1\}$. We shall see in this section that example (5.16) is no exception and that the structure of the closure of the "unit ball" of a bundle determines the strength of topological separation in the bundle space.

This section will contain a lot of rather technical results and I can only hope that the readers will not loose their patience before they reach the applications of the material presented here in sections 17 and 19.

16.1 Lemma. *Let* $p : E \to X$ *be a bundle with seminorms* $(\nu_j)_{j\in J}$. *Define* $\tilde{\nu}_j : E \to \mathbb{R}$ *by*

(1) $$\tilde{\nu}_j(\alpha) := \sup \{ r \in \mathbb{R} : \nu_j^{-1}(]r,\infty[) \text{ is a neighborhood of } \alpha\}.$$

Then $\tilde{\nu}_j$ *satisfies* $\tilde{\nu}_j(\lambda\cdot\alpha) = |\lambda|\cdot\tilde{\nu}_j(\alpha)$ *for all* $\lambda \in \mathbb{K}$.
Moreover, if $\sigma : U \to E$ *is a section where* $U \subset X$ *is open, then for every* $x \in U$ *we have*

(2) $$\tilde{\nu}_j(\sigma(x)) = \sup \{r \in \mathbb{R} : \{y : r < \nu_j(\sigma(y))\} \text{ is a neighborhood of } x\}$$

Furthermore

(3) $$\tilde{\nu}_j(\alpha) = \sup \{\inf \nu_j(\mathcal{U}) : \mathcal{U} \text{ is a neighborhood of } \alpha\}$$

Especially, $\nu_j(\alpha) > 0$ *for some* $j \in J$ *if and only if* α *and* $0 \in E_{p(\alpha)}$ *have disjoint neighborhoods in* E.

Proof. Firstly, let us check the last equality: As $\inf \nu_j(\nu_j^{-1}(]r,\infty[)) \geq r$, the right hand side of this equation dominates $\tilde{\nu}_j(\alpha)$. Conversely, let $\alpha \in E$ and let U be a neighborhood of α. Then for every $r < \inf \nu_j(U)$ the set $\nu_j^{-1}(]r,\infty[)$ contains U and thus is a neighborhood of α. This shows equality.

Now suppose that $\tilde{\nu}_j(\alpha) = \varepsilon > 0$. By the last equation there is a neighborhood U of α such that $\inf \nu_j(U) > \varepsilon/2$. Thus, U and $\{\beta \in E : \nu_j(\beta) < \varepsilon/2\}$ are disjoint neighborhoods of α and $0 \in E_{p(\alpha)}$, respectively.

Conversely, suppose that U and V are disjoint neighborhoods of α and $0 \in E_{p(\alpha)}$. We may assume w.o.l.g. that $V = \{\beta \in E : p(\beta) \in W$ and $\nu_j(\beta) < \varepsilon\}$ for some open set $W \subset X$, some $j \in J$ and some $\varepsilon > 0$. But then we have $\inf \nu_j(U \cap p^{-1}(W)) \geq \varepsilon$, i.e. $\tilde{\nu}_j(\alpha) \geq \varepsilon$. □

16.2 Lemma. *Let* $p : E \to X$ *be a bundle with seminorms* $(\nu_j)_{j\in J}$.

(i) The mappings $(\tilde{\nu}_j)_{j\in J}$ *are lower semicontinuous.*

(ii) Let $x_0 \in X$. *Then the restriction of* $\tilde{\nu}_j$ *to* E_{x_0} *is uniformly continuous for every* $j \in J$.

Proof. It remains to check (ii). Let $\varepsilon > 0$ and choose $\delta = \varepsilon/3$. Now assume that $\nu_j(\alpha - \beta) < \delta$. We show that $|\tilde{\nu}_j(\alpha) - \tilde{\nu}_j(\beta)| < \varepsilon$: Let $U = \{\gamma \in E : \nu_j(\gamma) < \delta\}$ and let V be any open neighborhood of α contained in $\{\gamma \in E : \nu_j(\gamma) > \tilde{\nu}_j(\alpha) - \delta\}$. As $\beta - \alpha \in U$ and as the mapping add : $E \vee E \to E$ is open by (1.6.(xi)), the set $W = \mathrm{add}(U,V) = \{\gamma_1 + \gamma_2 : p(\gamma_1) = p(\gamma_2),\ \gamma_1 \in U,\ \gamma_2 \in V\}$ is an open neighborhood of β. Moreover, if $\gamma \in W$, then $\gamma = \gamma_1 + \gamma_2$ for certain $\gamma_1 \in U$, $\gamma_2 \in V$. Thus the triangle inequality yields $\nu_j(\gamma) \geq \nu_j(\gamma_2) - \nu_j(\gamma_1) > \tilde{\nu}_j(\alpha) - \delta - \delta = \tilde{\nu}_j(\alpha) - 2\cdot\delta$. Thus, the set

$$\{\gamma \in E : \nu_j(\gamma) > \tilde{\nu}_j(\alpha) - 2\cdot\delta\}$$

is an open neighborhood of β. This yields the inequality

$\tilde{\nu}_j(\beta) \geq \tilde{\nu}_j(\alpha) - 2\cdot\delta$, i.e. $\tilde{\nu}_j(\alpha) - \tilde{\nu}_j(\beta) < \varepsilon$. By symmetry we get $\tilde{\nu}_j(\beta) - \tilde{\nu}_j(\alpha) < \varepsilon$ and therefore $|\tilde{\nu}_j(\alpha) - \tilde{\nu}_j(\beta)| < \varepsilon$. □

The following example shows that the mappings $(\tilde{\nu}_j)_{j\in J}$ are in general not continuous:

16.3 Example. Let $X = [-1,1] \subset \mathbb{R}$, equipped with its usual topology. We consider the following weight function $\omega : X \to \mathbb{R}$:

$$\omega(r) = \begin{cases} 1, & r \leq 0 \\ r, & r > 0 \end{cases}$$

and equip $C([-1,1])$ with the weighted norm $|||\cdot|||$ given by

$$\nu(f) = |||f||| = \sup\,\{\omega(r)\cdot|f(r)| : -1 \leq r \leq 1\}$$

Let E be the completion of $C([-1,1])$ in the norm $|||\cdot|||$. As in (5.16) we see that there is a bundle $p : E \to [-1,1]$ such that E is isometrically isomorphic to $\Gamma(p)$. It turns out that

$$\tilde{\nu}(\alpha) = \begin{cases} |||\alpha||| & , \ p(\alpha) \neq 0 \\ 0 & , \ p(\alpha) = 0 \end{cases}$$

Now assume that $\tilde{\nu}$ is continuous. Then for all $\sigma \in \Gamma(p)$ the composition $\tilde{\nu}\circ\sigma$ is continuous. But this is impossible, as by construction the constant mapping with value 1 belongs to $\Gamma(p)$.

A later example will show that the $\tilde{\nu}_j$ do not have to be seminorms, which in our situation means that they need not be sublinear.

16.4 Proposition. *Let $p : E \to X$ be a bundle and assume that X is Hausdorff. Then the following conditions are equivalent:*

(i) E is Hausdorff

(ii) $\{0_x \in E_x : x \in X\} \subset E$ is closed.

(iii) For every $0 \neq \alpha \in E$ there is a seminorm $\nu_j : E \to \mathbb{R}$ of the bundle $p : E \to X$ such that $\tilde{\nu}_j(\alpha) > 0$.

If X *is in addition normal and second countable, if all stalks are complete and if the bundle has a countable family of seminorms (especially if* $p : E \to X$ *is a bundle of Banach spaces with a normal and second countable base space), then conditions* (i) - (iii) *are also equivalent to*

(iv) For all $\sigma \in \Gamma(p)$ the set $\{x \in X : \sigma(x) = 0\}$ is closed.

Proof. (i) → (ii) : Let α belong to the closure of $\{0_x \in E_x : x \in X\}$. Then there is a net $(x_i)_{i \in I}$ such that $\alpha = \lim_{i \in I} 0_{x_i}$. As p is continuous, we conclude that $p(\alpha) = \lim_{i \in I} p(0_{x_i}) = \lim_{i \in I} x_i$ and as the 0-section is continuous, we may write $0_{p(\alpha)} = \lim_{i \in I} 0_{x_i}$. Because E is a Hausdorff space, limits of nets are unique, if they exist. Hence we conclude that $\alpha = 0_{p(\alpha)}$.

(ii) → (iii) : Let $0 \neq \alpha \in E$. As the set $\{0_x \in E_x : x \in X\}$ is closed, we can find an open neighborhood $\mathcal{U}$ of α such that $\beta \neq 0$ for all $\beta \in \mathcal{U}$. We may assume without loss of generality that

$$\mathcal{U} = \{\beta \in E : p(\beta) \in U,\ \nu_j(\sigma(p(\beta)) - \beta) < \varepsilon\}$$

for a certain open neighborhood U of $p(\alpha)$, a certain seminorm $\nu_j : E \to \mathbb{R}$ of our bundle and a certain local section $\sigma : U \to E$. Now let

$$\delta = \frac{1}{2}\cdot(\varepsilon - \nu_j(\sigma(p(\alpha)) - \alpha)).$$

Then the smaller set

$$\mathcal{V} = \{\beta \in E : p(\beta) \in U,\ \nu_j(\sigma(p(\beta)) - \beta) < \varepsilon - \delta\}$$

is still an open neighborhood of α. Moreover, for all $\beta \in V$ we have $\nu_j(\beta) > \delta$, as $\nu_j(\beta) \leq \delta$ for an element $\beta \in V$ would imply

$$\begin{aligned} \nu_j(\sigma(p(\beta)) &= \nu_j(\sigma(p(\beta)) - O_{p(\beta)}) \\ &\leq \nu_j(\sigma(p(\beta)) - \beta) + \nu_j(\beta) \\ &< \varepsilon - \delta + \delta \\ &= \varepsilon . \end{aligned}$$

Thus $O_{p(\beta)} \in U$ contradicting the choice of U. We now conclude that $\tilde{\nu}_j(\alpha) \geq \delta$.

(iii) $\rightarrow$ (i) : Let $\alpha, \beta \in E$ be two distinct elements of E. We have to show that they have disjoint neighborhoods. This is obvious if $p(\alpha) \neq p(\beta)$, as in this case we may take disjoint open neighborhoods U and V of $p(\alpha)$ and $p(\beta)$ resp. Then $p^{-1}(U)$ and $p^{-1}(V)$ are disjoint open neighborhoods of α and β respectively.
Now suppose that $p(\alpha) = p(\beta)$. In this case $\alpha \neq \beta$ implies $\alpha - \beta \neq O$. Using (iii) we can find a seminorm $\nu_j : E \rightarrow \mathbb{R}$ such that $\tilde{\nu}_j(\alpha - \beta) > O$. Let $r := \frac{1}{2}\cdot\tilde{\nu}_j(\alpha - \beta)$. Then there is an open neighborhood U Of $\alpha - \beta$ such that $\nu_j(\gamma) > r$ for all $\gamma \in U$. As the mapping

$$\begin{aligned} E \vee E &\rightarrow E \\ (\gamma_1, \gamma_2) &\rightarrow \gamma_1 - \gamma_2 \end{aligned}$$

is continuous, there are open neighborhoods V and W of α and β resp. such that $\gamma_1 - \gamma_2 \in U$ for all $\gamma_1 \in V$, $\gamma_2 \in W$. These sets V and W are disjoint: Indeed, if $\gamma \in V \cap W$, then $O = \gamma - \gamma \in U$, contradicting $\nu_j(\gamma') > r$ for all $\gamma' \in U$.

Finally, suppose that $p : E \rightarrow X$ satisfies the additional properties listed in (16.4). Then obviously (i) implies (iv). Conversely, we shall show that (iv) implies (ii):

Let $\alpha \in E$ belong to the closure of $\{0_x \in E_x : x \in X\}$. Assume, if possible, that $0 \neq \alpha$.
By (2.9) there is a section $\sigma \in \Gamma(p)$ such that $\sigma(p(\alpha)) = \alpha$ and by (1.6.(vii)) the sets of the form

$$T(U,\sigma,\varepsilon,j) := \{\beta \in E : p(\beta) \in U,\ \nu_j(\sigma(p(\beta)) - \beta) < \varepsilon\}$$

form a neighborhood base of α, where U runs through all open neighborhoods of $p(\alpha)$, ν_j runs through all seminorms of the bundle and where ε ranges over all positive numbers. Obviously, we may restrict ourselves to a countable neighborhood base of $p(\alpha)$ and to real numbers of the form $\frac{1}{n}$, $n \in \mathbb{N}$. As the family of seminorms was countable, too, we conclude that α has a countable neighborhood base $(V_n)_{n\in\mathbb{N}}$ and we may assume that $V_{n+1} \subset V_n$ for all $n \in \mathbb{N}$. Moreover, the singleton $\{0_{p(\alpha)}\}$ is closed and $\alpha \neq 0_{p(\alpha)}$. Therefore we may assume that $0_{p(\alpha)} \notin V_n$ for all $n \in \mathbb{N}$.
As α belongs to the closure of $\{0_x \in E_x : x \in X\}$, for every $n \in \mathbb{N}$ we can pick an element $x_n \in X$ such that $0_{x_n} \in V_n$. Obviously, $x_n \neq p(\alpha)$ for all $n \in \mathbb{N}$, $\lim_{n\to\infty} 0_{x_n} = \alpha$ and thus $\lim_{n\to\infty} x_n = p(\alpha)$. Hence the set

$$A := \{p(\alpha)\} \cup \{x_n : n \in \mathbb{N}\}$$

is compact and the selection $\rho : A \to E$ defined by

$$\rho(x) = \begin{cases} 0_{x_n} & \text{if } x = x_n \text{ for some } n \in \mathbb{N} \\ \alpha & \text{if } x = p(\alpha) \end{cases}$$

is continuous. By (4.4) we can find an extension $\bar{\rho} \in \Gamma(p)$ of ρ. For this section, the set $\{x : \bar{\rho}(x) = 0\}$ is not closed, as $\bar{\rho}(x_n) = \rho(x_n) = 0_{x_n}$, but $\bar{\rho}(\lim_n x_n) = \bar{\rho}(p(\alpha)) = \alpha \neq 0$. □

16.5 Corollary. *Let* $p : E \to X$ *be a bundle of finite dimensional spaces over a Hausdorff space* X. *Then* E *is Hausdorff if and only if for every* $x \in X$ *there is a seminorm* $\nu_{j_x} : E \to \mathbb{R}$ *and a real number*

$M_x > 0$ *such that for all* $0 \neq \alpha \in E_x$ *the set*

$$\{\beta \in E : \nu_{j_x}(\beta) > M_x \cdot \nu_{j_x}(\alpha)\}$$

is a neighborhood of α. *Moreover, in this case the seminorm* $\nu_{j_x} : E \to \mathbb{R}$ *may be chosen to be a norm when restricted to the stalk* E_x.

Proof. Assume that E is Hausdorff. As E_x is finite dimensional, the topology of E_x is induced by a norm $\|\cdot\|$. Because the sets of the form $\{\alpha \in E_x : \nu_j(\alpha) < \varepsilon\}$ form a (directed) neighborhood base at $0 \in E_x$, where $\nu_j : E \to \mathbb{R}$ runs through all seminorms of the bundle and where ε ranges over all positive numbers, we can find an index $j_o \in J$ and an $\varepsilon > 0$ such that $\{\alpha \in E_x : \nu_{j_o}(\alpha) < \varepsilon\} \subset \{\alpha \in E_x : \|\alpha\| < 1\}$. Hence the set $\{\alpha \in E_x : \nu_{j_o}(\alpha) < \varepsilon\}$ contains no non-trivial subspace of E_x and therefore ν_{j_o} is a norm on E_x. This also implies that ν_j is a norm on E_x whenever $\nu_{j_o} \leq \nu_j$. Moreover, the definition of the $\tilde{\nu}_j$ implies that $\tilde{\nu}_j \leq \tilde{\nu}_{j'}$ whenever $\nu_j \leq \nu_{j'}$.

Now we note that $B = \{\alpha \in E_x : \|\alpha\| = 1\}$ is compact. As E is Hausdorff, we can use (16.4) to find for a given element $\alpha \in B$ a seminorm $\nu_{j_\alpha} : E \to \mathbb{R}$ of the bundle such that $\tilde{\nu}_{j_\alpha}(\alpha) > 0$. By the above remark we may assume that $\nu_{j_o} \leq \nu_{j_\alpha}$. Now the sets of the form

$$U_\alpha = \{\beta \in E_x : \tilde{\nu}_{j_\alpha}(\beta) > 0\}$$

are open by the lower semicontinuity of the $\tilde{\nu}_j$. Since these sets cover the compact set B, we can pick elements $\alpha_1, \ldots, \alpha_n \in B$ such that $B \subset U_{\alpha_1} \cup \ldots \cup U_{\alpha_n}$. Choose any index j_x such that

$$\nu_{j_{\alpha_i}} \leq \nu_{j_x} \qquad \text{for all } i \in \{1, \ldots, n\}.$$

Then we conclude that $\tilde{\nu}_{j_x}(\alpha) > 0$ for all $\alpha \in B$. Moreover, ν_{j_x} is a norm on E_x as $\nu_{j_o} \leq \nu_{j_x}$. As the restriction of ν_{j_x} to E_x is continuous by (16.2), the set $\tilde{\nu}_{j_x}(\mathcal{B}) \subset \mathbb{R}$ is compact and does not contain 0. Hence we can find an $L_x > 0$ such that $\tilde{\nu}_{j_x}(\mathcal{B}) \subset]L_x,\infty[$. Now assume that $0 \neq \alpha \in E_x$. Then the element $\alpha/\|\alpha\|$ belongs to $\mathcal{B}$ and therefore we have $\tilde{\nu}_{j_x}(\alpha) > L_x \cdot \|\alpha\|$.

As every two norms on a finite dimensional space are equivalent, we can find a constant $M_x > 0$ such that $\|\alpha\| \cdot L_x \geq M_x \cdot \nu_{j_x}(\alpha)$ for all $\alpha \in E_x$. This implies $\tilde{\nu}_{j_x}(\alpha) > M_x \cdot \nu_{j_x}(\alpha)$ for all $\alpha \in E_x$ which are different from 0. In particular, by the definition of $\tilde{\nu}_{j_x}$ we conclude that the set $\{\beta \in E : \nu_{j_x}(\beta) > M_x \cdot \nu_{j_x}(\alpha)\}$ is a neighborhood of α for every $0 \neq \alpha \in E_x$. □

We now turn our attention to a different description of the ν_j:

16.6 Definition. Let $p : E \to X$ be a bundle with seminorms $(\nu_j)_{j \in J}$. For every $j \in J$ and every $M \in \mathbb{R}$ we define

$$C_{j,M} := \{\alpha \in E : \nu_j(\alpha) \leq M\}^-,$$

where $^-$ denotes the topological closure in E.

If $p : E \to X$ is a bundle of Banach spaces, we let

$$C_M := \{\alpha \in E : \|\alpha\| \leq M\}^-. \qquad \square$$

16.7 Proposition. *For every bundle* $p : E \to X$ *we have:*

(i) $\alpha \in C_{j,M}$ *if and only if* $\tilde{\nu}_j(\alpha) \leq M$, *provided that* $M \neq 0$.

(ii) $\tilde{\nu}_j(\alpha) = \inf \{M : \alpha \in C_{j,M}\}$

(iii) *If* $\sigma : U \to E$ *is a local section, then* $\tilde{\nu}_j(\sigma(x)) \leq M$ *if and only if* $x \in \bigcap_{\varepsilon > 0} \{y : \nu_j(\sigma(y)) \leq M + \varepsilon\}^-$.

(iv) *If* $p : E \to X$ *is a bundle of Banach spaces, then* (i) *holds also for* $M = 0$.

Proof. (i) : As $\tilde{\nu}_j$ is lower semicontinuous, the set $\tilde{\nu}_j^{-1}((-\infty,M])$ is

closed. As $\nu_j(\alpha) \le M$ implies $\tilde{\nu}_j(\alpha) \le M$, we obtain $\{\alpha \in E : \nu_j(\alpha) \le M\}$ $\subset \tilde{\nu}_j^{-1}((-\infty,M])$ and thus $C_{j,M} \subset \tilde{\nu}_j^{-1}((-\infty,M)$.
Conversely, suppose that $\tilde{\nu}_j(\alpha) \le M \ne 0$. Then for every $\varepsilon > 0$ we have $\tilde{\nu}_j(\alpha) < M + \varepsilon$. From the last equation in (16.1) we conclude that every neighborhood U of α contains an element β such that $\nu_j(\beta) <$ $< M + \varepsilon$ and from the definition of $C_{j,M}$ we deduce that $\alpha \in C_{j,M+\varepsilon}$ for every $\varepsilon > 0$. Since multiplication with scalars different from 0 is a homeomorphism, this yields $\frac{M}{M+\varepsilon}\cdot\alpha \in C_{j,M}$ for every $\varepsilon > 0$. Letting ε go to 0, we obtain $\alpha \in C_{j,M}$.

(ii): If $\tilde{\nu}_j(\alpha) \ne 0$, then (ii) follows immediately from (i). On the other hand, if $\tilde{\nu}_j(\alpha) = 0$, then using the same arguments as above, we may conclude that α belongs to $C_{j,\varepsilon}$ for every $\varepsilon > 0$.

(iii): Assume that x belongs to $\{y : \nu_j(\sigma(y)) \le M + \varepsilon\}^-$ for every $\varepsilon > 0$. Then for every $\varepsilon > 0$ we know that

$$\begin{aligned}\sigma(x) &\in \sigma(\{y : \nu_j(\sigma(y)) \le M + \varepsilon\}^-) \\ &\subset \sigma(\{y : \nu_j(\sigma(y)) \le M + \varepsilon\})^- \\ &\subset C_{j,M+\varepsilon}\end{aligned}$$

and thus $\tilde{\nu}_j(\sigma(x)) \le M + \varepsilon$ by (i). As $\varepsilon > 0$ was arbitrary, we obtain $\tilde{\nu}_j(\sigma(x)) \le M$.
Conversely, assume that $\tilde{\nu}_j(\sigma(x)) \le M$. Using (i) again, we conclude that $\sigma(x) \in \bigcap_{\varepsilon>0} \{\alpha \in E : \nu_j(\alpha) \le M + \varepsilon\}^-$. Suppose that the element x belongs to the open set $V := U \setminus \{y : \nu_j(\sigma(y)) \le M + \varepsilon\}^-$ for some $\varepsilon > 0$. Then for all $y \in V$ we have $\nu_j(\sigma(y)) > M + \varepsilon$ and therefore $\tilde{\nu}_j(\sigma(x)) \ge M + \varepsilon$ by (16.1), which is impossible. This concludes the proof of (iii).

(iv): Finally, let us suppose that $p : E \to X$ is a bundle of Banach spaces. We have to show that α belongs to the closure of $\{0_x \in E_x : x \in X\}$ if and only if $\tilde{\nu}(\alpha) = 0$, where $\nu : E \to \mathbb{R}$ is given by $\nu(\alpha) = \|\alpha\|$.

Assume that $\tilde{\nu}(\alpha) = 0$. If $\{\beta \in E : \|\beta\| > 0\}$ were a neighborhood of α, we could find a local section $\sigma : U \to E$ and an $\varepsilon > 0$ such that $\alpha \in \{\beta \in E : p(\beta) \in U, \|\sigma(p(\beta)) - \beta\| < \varepsilon\} \subset \{\beta \in E : \beta \neq 0\}$. Let $\delta = \frac{1}{2}\cdot(\varepsilon - \|\sigma(p(\alpha)) - \alpha\|)$. Then as in the proof of (16.4,(ii)→(iii)) we would see that $\tilde{\nu}(\alpha) \geq \delta > 0$, a contradiction.

This other implication holds trivially. □

The following example shows that the mapping $\nu_j : E \to \mathbb{R}$ need not be seminorms. To verify this, we shall construct a bundle of Banach spaces $p : E \to X$ such that the closure of the unit ball $\{\alpha \in E : \|\alpha\| \leq 1\}$ is not stalkwise convex and then apply (16.7.(i)).

16.8 Example. Let $X = [0,1]$ be the unit interval with the usual topology, let $E = [0,1] \times \mathbb{R}^2$, equipped with the product topology and let $p : E \to X$ be the first projection. We define a norm on E by

$$\|(r,(a,b))\| = \begin{cases} \max\{|a|/2, |b|\} , & r < 1/2 \\ \max\{|a|, |b|\} , & r = 1/2 \\ \max\{|a|, |b|/2\} , & r > 1/2 \end{cases}$$

Using (3.2), it is easy to check that $p : E \to X$ equipped with this norm is a bundle of Banach spaces.

Moreover, we have

$$\begin{aligned} C_1 \cap p^{-1}(\tfrac{1}{2}) &= \{\alpha \in E : \|\alpha\| \leq 1\}^- \cap p^{-1}(\tfrac{1}{2}) \\ &= \{(\tfrac{1}{2},a,b) : (|a| \leq 1 \text{ and } |b| \leq 2) \text{ or } \\ &\qquad (|a| \leq 2 \text{ and } |b| \leq 1)\} \end{aligned}$$

and this set is not convex.

In the remainder of this section we restrict our attention to bundles of Banach spaces. We shall continue the discussion of duality already begun in section 15.

Let us recall some notations ($p : E \to X$ is a bundle of Banach spaces with a compact base space):

B_x denotes the dual unit ball of the stalk E_x, identified with a subset of $\Gamma(p)'$ via the natural embedding $E'_x \to \Gamma(p)'$.

$B_A = \bigcup_{x \in A} B_x$, where A is a subset of X

The following result is a generalization of (15.9):

16.9 Proposition. *Let $p : E \to X$ be a bundle of Banach spaces over a compact base space X. For every $x \in X$ let $W_x \subset B_x$ be a closed and convex subset. If $A \subset X$ is a subset, then we define $W_A := \bigcup_{x \in A} W_x$. Then the following conditions are equivalent:*

(i) If $A \subset X$ is any subset, then $x \in \overline{A}$ implies $W_x \subset \overline{W_A}$.

(ii) If $A \subset X$ is any subset, then $x \in \overline{A}$ implies $W_x \subset \overline{\mathrm{conv}}\, W_A$.

(iii) If u is an ultrafilter on X with $\lim u = x$, then $W_x \subset \lim_u W_y$.

(iv) If $U \subset B_X$ is (relatively) open, then $\{x \in X : W_x \cap U \neq \emptyset\}$ is open in X.

(v) If $A \subset B_X$ is closed, then $\{x \in X : W_x \subset A\}$ is closed in X.

Here, the limit $\lim_u W_y$ is taken in the Lawson topology of Cl B_X.

Proof. (i) $\to$ (ii) : is trivial.

(ii) $\to$ (iii): As all the sets W_x , $x \in X$, are closed and convex and as the embedding Conv $\mathcal{B}_X \to \mathrm{Cl}\ \mathcal{B}_X$ is continuous by (12.7), we have $\lim_{\mathcal{U}} \mathcal{W}_y = \cap \{\overline{\mathrm{conv}\ \mathcal{B}_A} : A \in \mathcal{U}\}$ by (12.8). As $A \in \mathcal{U}$ implies $x \in \overline{A}$, the property (iii) follows from (ii).

(iii) $\to$ (v): Let $A \subset \mathcal{B}_X$ be closed and let $\mathcal{U}$ be an ultrafilter containing $\{x \in X : \mathcal{W}_x \subset A\}$. We have to show that $\lim \mathcal{U} \in \{x : \mathcal{W}_x \subset A\}$, i.e. $\mathcal{W}_{\lim \mathcal{U}} \in A$. But this is true, since we only have to note that $\{B \in \mathrm{Cl}\ \mathcal{B}_X : B \subset A\}$ is closed in the Lawson topology of $\mathrm{Cl}\ \mathcal{B}_X$ and as $\{x \in X : \mathcal{W}_x \subset A\} \in \mathcal{U}$. Therefore, using (iii) we may conclude that $\mathcal{W}_{\lim \mathcal{U}} \subset \lim_{\mathcal{U}} \mathcal{W}_y \subset A$.

(iv) $\to$ (v) and (v) $\to$ (iv) are trivial.

(v) $\to$ (i): Let $x \in \overline{A}$. Then by (v) the set $\{y : \mathcal{W}_y \subset \overline{\mathcal{W}_A}\}$ is closed. As this set contains A, it contains the closure of A, too. This yields $\mathcal{W}_x \subset \overline{\mathcal{W}_A}$. □

The next result allows us to use duality in order to identify all stalkwise convex and closed subsets of E containing the unit ball $\{\alpha \in E : \|\alpha\| \leq 1\}$:

16.10. Proposition. *Let* $p : E \to X$ *be a bundle of Banach spaces with a compact base space* X. *Then a subset* $C \subset E$ *containing* $\{\alpha \in E : \|\alpha\| \leq 1\}$ *such that* $C \cap E_x$ *is convex for every* $x \in X$ *is closed if and only if the family* $\mathcal{W}_x = (C \cap E_x)^{\circ} \subset \mathcal{B}_x$ *satisfies the equivalent conditions of* (16.9).

Proof. Let $C \subset E$ be given and let us assume that C is closed, stalkwise convex and contains $\{\alpha \in E : \|\alpha\| \leq 1\}$. Moreover, let A be a subset of X and let x_o belong to the closure of A. We have to show that $W_{x_o} \subset \overline{\text{conv}}\, W_A = W_A^{oo}$. Let us compute:

$$\begin{aligned} W_A^o &= (\bigcup_{x\in A} W_x)^o \\ &= \bigcap_{x\in A} W_x^o \\ &= \bigcap_{x\in A} \{\sigma \in \Gamma(p) : \sigma(x) \in C\} \\ &= \{\sigma \in \Gamma(p) : \sigma(A) \subset C\} \\ &= \{\sigma \in \Gamma(p) : \sigma(\overline{A}) \subset C\} \qquad \text{(since C is closed)} \\ &\subset \{\sigma \in \Gamma(p) : \sigma(x_o) \subset C\}. \end{aligned}$$

Now the result follows immediately by taking polars.

Conversely, assume that the family $W_x = (C \cap E_x)^o$, $x \in X$, satisfies condition (v) of (16.9). We have to show that C is closed, or, equivalently, that $E \setminus C$ is open. Thus, let us start with an element $\alpha \in E \setminus C$, let $x_o = p(\alpha)$ and choose $\sigma \in \Gamma(p)$ such that $\sigma(x_o) = \alpha$. For each $\varepsilon \geq 0$ we define

$$A_\varepsilon = \{\phi \in B_X : \text{Re}\, \phi(\sigma) \leq 1 + \varepsilon\}.$$

Then all the sets A_ε are closed and

$$A_o = \bigcap_{\varepsilon>0} A_\varepsilon .$$

If $W_{x_o} = (C \cap E_{x_o})^o$ were contained in A_ε for every $\varepsilon > 0$, then it would be contained in A_o, too. But for each $x \in X$ we have $(C \cap E_x)^o = W_x \subset A_o = \{\sigma\}^o$ if and only if $\sigma \in (C \cap E_x)^{oo} = \{\rho \in \Gamma(p) : \rho(x) \in C\}$. This means that $W_x \subset A_o$ if and only if $\sigma(x) \in C$. As $\alpha = \sigma(x_o)$ was not in C, we conclude that $W_{x_o} \not\subset A_o$.
Let $\varepsilon > 0$ be a positive real number such that $W_{x_o} \not\subset A_\varepsilon$ and let

$$U := \{x \in X : W_x \cap (B_X \setminus A_\varepsilon) \neq \emptyset\}.$$

Then $x_o \in U$ and U is open by (16.9.(v)). We define

$$\mathcal{O} := \{\beta \in \mathcal{E} : p(\beta) \in U \text{ and } \|\sigma(p(\beta)) - \beta\| < \varepsilon\}.$$

Then $\mathcal{O}$ is an open neighborhood of α. Moreover, $\mathcal{O} \cap C = \emptyset$: Indeed, for $\beta \in C \cap \mathcal{O}$ we would conclude that $p(\beta) \in U$, $\|\sigma(p(\beta)) - \beta\| < \varepsilon$ and $\mathrm{Re}\,\phi(\beta) \leq 1$ for all $\phi \in \mathcal{W}_{p(\beta)} = (C \cap E_{p(\beta)})^{o}$. For all $\phi \in \mathcal{W}_{p(\beta)}$ this would imply the inequality

$$\begin{aligned}
\mathrm{Re}\,\phi(\sigma(p(\beta))) &= \mathrm{Re}\,(\phi(\beta) + (\phi(\sigma(p(\beta))) - \phi(\beta))) \\
&= \mathrm{Re}\,\phi(\beta) + \mathrm{Re}\,(\phi(\sigma(p(\beta))) - \phi(\beta)) \\
&\leq \mathrm{Re}\,\phi(\beta) + |\phi(\sigma(p(\beta))) - \phi(\beta)| \\
&\leq \mathrm{Re}\,\phi(\beta) + \|\phi\| \cdot \|\sigma(p(\beta)) - \beta\| \\
&\leq 1 + \varepsilon
\end{aligned}$$

and thus $\mathcal{W}_{p(\beta)} \subset A_\varepsilon$ contradicting the fact that $p(\beta) \in U =$ $= \{x : \mathcal{W}_x \cap (\mathcal{B}_x \setminus A_\varepsilon) \neq \emptyset\}$. □

In a later section, when we shall talk about the "internal" dual of all $\Gamma(p)$ consisting of all $C(X)$-module homomorphisms from $\Gamma(p)$ into $C(X)$, we shall need the largest family of subsets $\mathcal{W}_x \subset \mathcal{B}_x$ such that the properties of (16.9) are satisfied. By (16.10) we know that this family is determined by the smallest stalkwise convex, closed subset C of $\mathcal{E}$ containing $\{\alpha : \|\alpha\| \leq 1\}$. Hence it seems to be desirable to have an explicit description of this set. I conjecture that this set can be obtained by taking stalkwise the closed convex hull of the topological closure of $\{\alpha \in \mathcal{E} : \|\alpha\| \leq 1\}$, but I do not have any proof hereof. All I am able to do is to identify the stalkwise polars of this set:

16.11 Notation. Let $p : \mathcal{E} \to X$ be a bundle of Banach spaces, X compact. For every $x \in X$ we define a subset $\mathcal{K}_x \subset \mathcal{B}_x$ by

$$K_x := \cap \{\overline{B_M} : M \subset X \text{ and } x \in \overline{M}\}.$$

16.12 Proposition. *Let* $p : E \to X$ *be a bundle of Banach spaces with a compact base space* X. *Then* K_x *is circled and*

(i) $K_x \subset B_x$

(ii) $K_x = \cap \{B_M^{OO} : x \in \overline{M}\} = \cap \{\overline{\text{conv}} \; B_M : x \in \overline{M}\} = \cap \{\lim_{\mathcal{u}} B_x :$
       $\mathcal{u}$ *is an ultrafilter on X converging to x*$\}$

where $\lim_{\mathcal{u}} B_x$ *is calculated in the Lawson topology on* Cl B_X.

Proof. As B_M is always circled, so is $\overline{B_M}$. Since K_x is an intersection of sets of this form, K_x is circled, too.

(i) : From $x \in \overline{\{x\}}$, we have $K_x \subset B_x$.

(ii) : For every subset $M \subset X$ we have $\overline{B_M} \subset B_M^{OO}$ and therefore $K_x \subset \cap \{B_M^{OO} : x \in \overline{M}\} \subset \cap \{\overline{\text{conv}} \; B_M : x \in \overline{M}\}$.

Moreover, let $\mathcal{u}$ be an ultrafilter on X converging to x and let $B_{\mathcal{u}}$ be the ultrafilter generated by the image of $\mathcal{u}$ under the mapping $x \to B_x : X \to \text{Cl } B_1(\Gamma(p)')$ (recall that $B_1(\Gamma(p)')$ denotes the unit ball of $\Gamma(p)'$). Then $\lim B_{\mathcal{u}} = \lim_{\mathcal{u}} B_x$ and $B_{\mathcal{u}}$ is an ultrafilter having a base whose elements consist of closed convex sets only. Hence, by (12.7), the limits of $B_{\mathcal{u}}$ in Cl $B_1(\Gamma(p)')$ and Conv $B_1(\Gamma(p)')$ agree. We may now calculate

$$\lim B_{\mathcal{u}} = \bigcap_{M \in \mathcal{u}} \overline{(\bigcup_{y \in M} B_y)}$$

$$= \bigcap_{M \in \mathcal{u}} \overline{B_M}$$

$$= \bigcap_{M \in \mathcal{u}} B_M^{OO}.$$

As $\lim \mathcal{u} = x$ implies $x \in \overline{M}$ for all $M \in \mathcal{u}$, we conclude that

$$\bigcap_{x \in \overline{M}} B_M^{OO} \subset \lim_{\mathcal{u}} B_{\mathcal{u}}$$

for every ultrafilter $\mathcal{u}$ with $\lim \mathcal{u} = x$. We obtain the inequality

$\cap\, \{\mathcal{B}_M^{oo} : x \in \overline{M}\} \subset \cap\, \{\lim_{\mathcal{U}} \mathcal{B}_x : \mathcal{U}$ is an ultrafilter on X converging to $x\}$.

It remains to show that the last set is contained in K_x. To do so, we have to prove that for every $M \subset X$ with $x \in \overline{M}$ there is an ultrafilter $\mathcal{U}$ on X converging to x such that $\lim_{\mathcal{U}} \mathcal{B}_y \subset \overline{\mathcal{B}_M}$. But this is easy, as $x \in \overline{M}$ implies the existence of an ultrafilter $\mathcal{U}$ on X with $M \in \mathcal{U}$ and $\lim \mathcal{U} = x$. For any such ultrafilter, we have

$$\begin{aligned} \lim_{\mathcal{U}} \mathcal{B}_y &= \bigcap_{N \in \mathcal{U}} \overline{(\bigcup_{y \in M} \mathcal{B}_y)} \\ &= \bigcap_{N \in \mathcal{U}} \overline{\mathcal{B}_M} \\ &\subset \overline{\mathcal{B}_M}. \qquad \square \end{aligned}$$

16.13 Proposition. *Let* $p : E \to X$ *be a bundle of Banach spaces with a compact base space* X. *Then for every* $x \in X$ *we have*

$$K_x = (C_1 \cap E_x)^o.$$

Proof. Let $\phi \in (C_1 \cap E_x)^o$. We have to show that ϕ belongs to $\mathcal{B}_M^{oo}$ for every subset $M \subset X$ with $x \in \overline{M}$. Firstly, note that σ belongs to $\mathcal{B}_M^o$ if and only if $\|\sigma(y)\| \leq 1$ for all $y \in M$. Hence, $x \in \overline{M}$ and $\sigma \in \mathcal{B}_M^o$ imply $\sigma(x) \in \sigma(\overline{M}) \subset \overline{\sigma(M)} \subset C_1$, i.e. $\sigma(x) \in C_1 \cap E_x$. Thus, we have shown that $(C_1 \cap E_x)^o \subset \mathcal{B}_M^{oo}$ whenever $x \in \overline{M}$, i.e. $(C_1 \cap E_x)^o \subset K_x$.

Conversely, assume that $\phi \in K_x$ and let $\alpha \in C_1 \cap E_x$. We have to show that $\operatorname{Re} \phi(\alpha) \leq 1$. As $C_1 \cap E_x$ is circled, $\operatorname{Re} \phi(\alpha) \leq 1$ holds if and only if $|\phi(\alpha)| \leq 1$.

Let $\sigma \in \Gamma(p)$ be a section such that $\sigma(x) = \alpha$ and let $\varepsilon > 0$. Using (16.7.(iii)), we conclude that x belongs to $\overline{M}$, where $M = \{y \in X : \|\sigma(y)\| \leq 1 + \varepsilon\}$, and therefore ϕ belongs to $\overline{\mathcal{B}_M}$. As for all $\psi \in \mathcal{B}_M$ we have $\psi(\sigma) = \psi(\sigma(y))$ for a certain $y \in M$, we may estimate

$$\begin{aligned} |\psi(\sigma)| &= |\psi(\sigma(y))| \\ &\leq \|\psi\| \cdot \|\sigma(y)\| \\ &\leq 1\cdot(1+\varepsilon) \; . \end{aligned}$$

As ϕ belongs to the closure of $\mathcal{B}_M$, this implies $|\phi(\sigma)| \leq 1 + \varepsilon$. Since $\varepsilon > 0$ was arbitrary, we conclude that $|\phi(\sigma)| \leq 1$. □

17. Locally trivial bundles: A definition

In this section we shall introduce locally trivial bundles, a classical concept which has been used in differential geometrie and algebraical topology since a long time. We should note however, that our definition of locally trivial bundles will differ slightly from the usual one, the reason being that homoeomorphisms between bundle spaces commuting with the projections are in general not what we call isomorphisms of bundles (see (10.1(i)d) and example (10.25). Nevertheless, if the base space is locally compact, our notion of locally trivial bundles will agree with the usual one.

17.1 Definition. (i) Let $p : E \to X$ and $q : F \to X$ be two bundles of Ω-spaces having the same base space X. We say that p and q are *locally isomorphic*, if every point $x \in X$ has a neighborhood $U_x \subset X$ such that the bundles $p : p^{-1}(U_x) \to U_x$ and $q : q^{-1}(U_x) \to U_x$ are isomorphic.

(ii) A bundle $p : E \to X$ is said to be *locally trivial* if it is locally isomorphic to a trivial bundle $pr_1 : X \times E \to X$, where E is a topological Ω-space (see example (1.8.(i)). The Ω-space E is called the *stalk* of the bundle $p : E \to X$. □

From (10.10) and (1.10) we conclude:

17.2 Proposition. *Let* $p : E \to X$ *be a bundle with a family of seminorms* $(\nu_j)_{j \in J}$.

(i) If p *is locally trivial, then every point* $x \in X$ *has a neighborhood* U_x *such that* $\Gamma_{U_x}(p)$ *and* $C_b(U_x, E)$ *are isomorphic (as* $C_b(U_x)$*-modules and as* Ω*-spaces, if required), where* E *is the stalk of the bundle* p.

(ii) If X *is completely regular and if the bundle* $p : E \to X$ *is full, then the converse also holds.* □

It is obvious that we have to insist on full bundles in order to get the converse of (17.2.(i)): By (2.3) fullness is a "local" property for bundles with a completely regular base space, and local properties are preserved under local isomorphy. As every trivial bundle is locally full, so is every locally trivial bundle. Hence every locally trivial bundle over a completely regular base space has to be full.

18. Local linear independence

Let us start with a locally trivial bundle $p : E \to X$ and let us suppose that we are given linear independent elements $\alpha_1,\ldots,\alpha_n \in E_x \subset E$. Then it is easy to see that we may find an open neighborhood U of x and sections $\sigma_1,\ldots,\sigma_n \in \Gamma_U(p)$ such that $\sigma_1(y),\ldots,\sigma_n(y)$ are linearly independent for every $y \in U$ and, moreover, $\sigma_i(x) = \alpha_i$ for every $1 \le i \le n$.

Unfortunately, as example (5.16) and the section $\chi_{\{0\}}$ defined there show, this property does not characterize locally trivial bundles. However, by adding some separations axioms to both E and X, this property gives us the right idea for such a characterization.

We use again the notations of section 16: If $p : E \to X$ is a bundle with seminorms $(\nu_j)_{j\in J}$, then we define as in (16.1)

$$\tilde{\nu}_j(\alpha) := \sup\{r \in \mathbb{R} : \{\beta : \nu_j(\beta) > r\} \text{ is a neighborhood of } \alpha\}.$$

Recall that $\tilde{\nu}_j(\alpha) > 0$ if and only if α and $0_{p(\alpha)}$ have disjoint neighborhoods.

The results of this sections are known for bundles of Banach spaces with continuous norm (see [Go 49]).

18.1 Proposition. *Let* $p : E \to X$ *be a bundle with seminorms* $(\nu_j)_{j\in J}$, *let* $x_o \in X$ *and let* V *be an open neighborhood of* x_o. *Assume that* $\sigma_1,\ldots,\sigma_n \in \Gamma_V(p)$ *are given such that* $\sigma_1(x_o),\ldots,\sigma_n(x_o)$ *are linearly independent. If for each* $0 \ne \alpha \in \langle\sigma_1(x_o),\ldots,\sigma_n(x_o)\rangle$ *in the linear span of the* $\sigma_i(x_o)$ *there is an index* $j \in J$ *such that* $\tilde{\nu}_j(\alpha) > 0$, *then there is a neighborhood* $U \subset V$ *of* x_o *such that for every* $x \in U$ *the set* $\{\sigma_1(x),\ldots,\sigma_n(x)\}$ *is linearly independent.*

Proof. Assume, if possible, that (18.1) is false. Then there is a net $(x_i)_{i\in I}$ converging to x_o and numbers $(r_{1,i})_{i\in I},\dots,(r_{n,i})_{i\in I}$ such that for every $i \in I$ we have

$$m_i := \max \{|r_{1,i}|,\dots,|r_{n,i}|\} > 0$$

and

$$\sum_{k=1}^{n} r_{,i}\cdot\sigma(x_i) = 0.$$

By dividing all the $r_{k,i}$ by m_i, we may assume that $|r_{k,i}| \le 1$ for all $i \in I$ and all $k \in \{1,\dots,n\}$. By multiplying with a unimodular number if necessary, we may assume that one of the $r_{k,i}$ is equal to 1 for all $i \in I$. Furthermore, there is an index $k_o \in \{1,\dots,n\}$ such that $I_o = \{i \in I : r_{k_o,i} = 1\}$ is cofinal in I; without loss of generality we may assume that $k_o = 1$. Hence, by substituting I by I_o, we may assume that $r_{1,i} = 1$ for all $i \in I$. Finally, by selecting a suitable subnet, we may assume that

$$r_k := \lim_{i\in I} r_{k,i}$$

exists for all $1 \le k \le n$.

Now $r_k\cdot\sigma_k(x_o)$ is a limit point of the net $(r_{k,i}\cdot\sigma_k(x_i))_{i\in I}$, as the scalar multiplication is continuous. As the addition is continuous, too, we conclude that $\sum_{k=1}^{n} r_k\cdot\sigma_k(x_o)$ is a limit point of $(\sum_{k=1}^{n} r_{k,i}\cdot\sigma_k(x_i))_{i\in I} = (0_{x_i})_{i\in I}$. Clearly, 0_{x_o} is also a limit point of the latter net. Now suppose that

$$\alpha := \sum_{k=1}^{n} r_k\cdot\sigma_k(x_o) \neq 0$$

Then we could find an $j \in J$ such that $\tilde{\nu}_j(\alpha) > 0$, which means that α and 0_{x_o} have disjoint neighborhoods. As they are both a limit point of the same net, this is impossible. Hence $\alpha = 0$, contradicting the fact that the $\sigma_k(x_o)$, $1 \le k \le n$, are linearly independent and that $r_1 = 1$. □

From the last proposition and (15.4) we conclude:

18.2 Theorem. *Let* $p : E \to X$ *be a bundle and assume that* E *is Hausdorff. Moreover, let* $\sigma_1,\ldots,\sigma_n \in \Gamma(p)$. *Then the set* $\{x \in X : \sigma_1(x),\ldots,\sigma_n(x)$ are linearly independent$\}$ *is open.* □

18.3 Theorem. *Let* $p : E \to X$ *be a bundle and assume that* E *is Hausdorff. Then the mapping*

$$\dim : X \to \mathbb{R}$$
$$x \to \dim E_x$$

is lower semicontinuous.

Proof. By definition (1.5) the set $\{\sigma(x) : \sigma \in \Gamma_U(p)$ for some open neighborhood U of $x\}$ is dense in E_x. Thus, if $\dim E_x \geq n$, then there are open neighborhoods $U_1,\ldots,U_n$ of x and sections $\tilde{\sigma}_i \in \Gamma_{U_i}(p)$, $1 \leq i \leq n$, such that the set $\{\tilde{\sigma}_1(x),\ldots,\tilde{\sigma}_n(x)\}$ is lineraly independent. Let $V := U_1 \cap \ldots \cap U_n$ and let $\sigma_i := \tilde{\sigma}_{i/V}$. Then (18.4) and (18.1) yield an open neighborhood $U \subset V$ of x such that for all $y \in U$ the set $\{\sigma_1(y),\ldots,\sigma_n(y)\}$ is linearly independent. Especially, we have $\dim E_y \geq n$ for all $y \in U$. □

We continue with a result which may be thought of as an improvement of (18.1):

18.4 Proposition. *Let* $p : E \to X$ *be a bundle with seminorms* $(\nu_j)_{j\in J}$, *let* $x_o \in X$ *be a point and let* V *be a neighborhood of* x_o. *Furthermore, let* $\sigma_1,\ldots,\sigma_n \in \Gamma_V(p)$ *be such that* $\sigma_1(x_o),\ldots,\sigma_n(x_o)$ *are linearly independent and assume that for every* $0 \neq \alpha \in \langle\sigma_1(x_o),\ldots$ $\ldots,\sigma_n(x_o)\rangle \subset E_{x_o}$ *there is an* $j \in J$ *such that* $\nu_j(\alpha) > 0$. *Then we can find an open neighborhood* $W \subset V$ *of* x_o *such that each neighborhood*

$U \subset W$ *of* x_o *has the following properties:*

(i) The $C_b(U)$*-submodule of* $\Gamma_U(p)$ *generated by* $\sigma_{1/U},\ldots,\sigma_{n/U} \in \Gamma_U(p)$ *is topologically and algebraically isomorphic to* $C_b(U, \mathbb{K}^n)$.

(ii) The $C_b(U)$*-submodule of* $\Gamma_U(p)$ *generated by* $\sigma_{1/U},\ldots,\sigma_{n/U} \in \Gamma_U(p)$ *is complete and hence closed in* $\Gamma_U(p)$.

Proof. Let $A \subset \mathbb{K}^n$ be defined by

$$A := \{(r_1,\ldots,r_n) \in \mathbb{K}^n : \max\{|r_1|,\ldots,|r_n|\} = 1\}$$

Then A is compact and therefore the set

$$\mathcal{A} := \{\sum_{i=1}^{n} r_i\cdot\sigma_i(x_o) : (r_1,\ldots,r_n) \in A\}$$

is compact in E_{x_o}. As the set $\{\sigma_1(x_o),\ldots,\sigma_n(x_o)\}$ is linearly independent and as $(0,\ldots,0) \notin A$, we conclude that $0 \notin \mathcal{A}$. Thus, for every $\alpha \in \mathcal{A}$ there is an index $j_\alpha \in J$ such that

$$0 < \varepsilon_\alpha := \tilde{\nu}_{j_\alpha}(\alpha).$$

Now by (16.2) the sets $\mathcal{O}_\alpha := \{\beta \in E_{x_o} : \tilde{\nu}_j(\beta) > \varepsilon_\alpha/2\}$ are open and cover $\mathcal{A}$. As $\mathcal{A}$ is compact, we can find $\alpha_1,\ldots,\alpha_n \in \mathcal{A}$ such that $\mathcal{A} \subset \mathcal{O}_{\alpha_1} \cup \ldots \cup \mathcal{O}_{\alpha_n}$. As the family of seminorms of a bundle is always directed, we can find an index $j_o \in J$ such that $\nu_j \geq \nu_{j_o}$ for all $j \in \{j_{\alpha_1},\ldots,j_{\alpha_n}\}$. Now define

$$\delta := \frac{1}{2}\cdot\min\{\varepsilon_{\alpha_1},\ldots,\varepsilon_{\alpha_n}\}.$$

Then it is easy to check that

$$\tilde{\nu}_{j_o}(\alpha) > \delta \qquad \text{for all } \alpha \in \mathcal{A}.$$

Now let $(r_{1,1},\ldots,r_{1,n}),(r_{2,1},\ldots,r_{2,n}),\ldots,(r_{m,1},\ldots,r_{m,n}) \in A$ be such that for each $(r_1,\ldots,r_n) \in A$ there is a certain $l \in \{1,\ldots,m\}$ with

$|r_1 - r_{1,1}| + \dots + |r_n - r_{1,n}| < \frac{\delta}{3 \cdot M}$, where

$$M := \max \{\sup_{y \in V} \nu_{j_o}(\sigma_1(y)), \dots, \sup_{y \in V} \nu_{j_o}(\sigma_n(y))\}.$$

Then we obtain the inequality:

$$\begin{aligned}
\sup_{y \in V} \nu_{j_o}\left(\sum_{i=1}^{n} (r_i \cdot \sigma_i(y) - r_{1,i} \cdot \sigma_i(y))\right) &\le \sum_{i=1}^{n} |r_i - r_{1,i}| \cdot \sup_{y\ V} \nu_{j_o}(\sigma_i(y)) \\
&\le \sum_{i=1}^{n} |r_i - r_{1,i}| \cdot M \\
&\le \delta/3.
\end{aligned}$$

Then the triangle inequality yields for all $y \in V$ the relation

$$\nu_{j_o}\left(\sum_{i=1}^{n} r_i \cdot \sigma_i(y)\right) \ge \nu_{j_o}\left(\sum_{i=1}^{n} r_{1,i} \cdot \sigma_i(y)\right) - \delta/3.$$

Now use (16.1) and (18.1) to find a neighborhood $W \subset V$ of x_o such that

(1) if $y \in W$, then the set $\{\sigma_1(y), \dots, \sigma_n(y)\}$ is linearly independent.

(2) $\nu_{j_o}\left(\sum_{i=1}^{n} r_{1,i} \cdot \sigma_i(y)\right) > \delta$ for all $y \in W$ and all $1 \in \{1, \dots, n\}$,

and let $U \subset W$ be any neighborhood of x_o. Then we conclude that

(1') if $y \in U$, then the set $\{\sigma_1(y), \dots, \sigma_n(y)\}$ is linearly independent.

(2') $\nu_{j_o}\left(\sum_{i=1}^{n} r_i \cdot \sigma_i(y)\right) > \frac{2}{3} \cdot \delta$ for all $(r_1, \dots, r_n) \in A$ and all $y \in U$.

Let $C_b(U)^n$ be equipped with the norm $\|f_1, \dots, f_n\| := \sum_{i=1}^{n} \|f_i\|$ and let $\Gamma_U(p)$ (as usual) be equipped with the family of seminorms $(\hat{\nu}_j)_{j \in J}$ given by $\hat{\nu}_j(\sigma) = \sup_{y \in U} \nu_j(\sigma(y))$. We define an operator

$$\begin{aligned}
T : \quad C_b(U)^n &\to \Gamma_U(p) \\
(f_1, \dots, f_n) &\to \sum_{i=1}^{n} f_i \cdot \sigma_i
\end{aligned}$$

As $C_b(U)^n \simeq C_b(U, \mathbb{K}^n)$, the proof will be complete if we can show that

T is a continuous and injective $C_b(U)$-module homomorphism which is open onto its image.

Obviously, T is a $C_b(U)$-module homomorphism.

To show the injectivity, let $T(f_1,\ldots,f_n) = 0$. Then for all $y \in U$ we have $\sum_{i=1}^{n} f_i(y)\cdot\sigma_i(y) = 0$ and thus $f_i(y) = 0$ for all $y \in U$ and all $i \in \{1,\ldots,n\}$ by (1'). This implies $(f_1,\ldots,f_n) = (0,\ldots,0)$.

Furthermore, T is continuous, as for all $(f_1,\ldots,f_n)$ and all $j \in J$ we have

$$\begin{aligned}\wp_j(T(f_1,\ldots,f_n)) &= \wp_j(\sum_{i=1}^{n} f_i\cdot\sigma_i|U)\\ &\le \sum_{i=1}^{n} \|f_i\|\cdot\wp_j(\sigma_i)\\ &\le \|(f_1,\ldots,f_n)\|\cdot\max\{\wp_j(\sigma_1),\ldots,\wp_j(\sigma_n)\}\end{aligned}$$

It remains to show that T is open onto its image:

Let $(f_1,\ldots,f_n) \in C_b(U)^n$ and let $\varepsilon > 0$. Then there is a $y \in U$ such that

$$\max\{\|f_1\|,\ldots,\|f_n\|\} \le \max\{|f_1(y)|,\ldots,|f_n(y)|\} + \frac{3\cdot\varepsilon}{2\cdot n\cdot\delta}$$

If we abbreviate $m = \max\{|f_1(y)|,\ldots,|f_n(y)|\}$, then we have $(\frac{1}{m}f_1(y),\ldots,\frac{1}{m}f_n(y)) \in A$ and therefore $\nu_{j_o}(\sum_{i=1}^{n}\frac{1}{m}f_i(y)\cdot\sigma_i(y)) > \frac{2}{3}\cdot\delta$ by (2'). We now have

$$\begin{aligned}\wp_{j_o}(T(f_1,\ldots,f_n)) &\ge \nu_{j_o}(T(f_1,\ldots,f_n)(y))\\ &= \nu_{j_o}(\sum_{i=1}^{n} f_i(y)\cdot\sigma_i(y))\\ &> m\cdot\frac{2}{3}\cdot\delta\end{aligned}$$

and this inequality yields

$$\frac{2}{3}\cdot\delta\cdot\|(f_1,\ldots,f_n)\| = \frac{2}{3}\cdot\delta\cdot\sum_{i=1}^{n}\|f_i\|$$

$$\leq \frac{2}{3}\cdot\delta\cdot n\cdot\max\{\|f_1\|,\ldots,\|f_n\|\}$$

$$\leq \frac{2}{3}\cdot\delta\cdot n\cdot(\max\{|f_1(y)|,\ldots,|f_n(y)|\} + \frac{3\cdot\varepsilon}{2\cdot n\cdot\delta})$$

$$= \frac{2}{3}\cdot\delta\cdot m\cdot n + \varepsilon$$

$$< n\cdot\Diamond_{j_o}(T(f_1,\ldots,f_n)) + \varepsilon$$

As $\varepsilon > 0$ was arbitrary, we conclude that $\frac{2\cdot\delta}{3\cdot n}\cdot\|(f_1,\ldots,f_n)\| \leq$ $\leq \Diamond_{j_o}(T(f_1,\ldots,f_n))$. Hence, T is open onto its image. □

18.5 Theorem. *Let* $p : E \to X$ *be a bundle with a locally compact base space* X. *Assume that all stalks of* p *have dimension* n, *where* $n \in \mathbb{N}$ *is fixed. Then the bundle* $p : E \to X$ *is locally trivial if and only if the bundle space* E *is Hausdorff.*

Proof. By definition, every locally trivial bundle over a Hausdorff base space has a Hausdorff bundle space.

Conversely, assume that E is Hausdorff and assume that all stalks have dimension n. Given a point $x_o \in X$, we have to find a neighborhood U of x such that the bundle $p_{/p^{-1}(U)} : p^{-1}(U) \to U$ and the trivial bundle $pr_1 : U \times \mathbb{K}^n \to U$ are isomorphic.

Let $\alpha_1,\ldots,\alpha_n$ be a base of E_{x_o}. As the bundle $p : E \to X$ is full by (2.12), there is a neighborhood W of x_o and sections $\sigma_1,\ldots,\sigma_n \in \Gamma_W(p)$ such that $\sigma_i(x_o) = \alpha_i$ for all $1 \leq i \leq n$. Applying (17.2) we can find a neighborhood $V \subset W$ such that $\{\sigma_1(y),\ldots,\sigma_n(y)\}$ is linearly independent and thus a base of E_y for every $y \in V$. Moreover, by (18.4) we can find a compact neighborhood $U \subset V$ of x_o such that the C(U)-submodule of $\Gamma_U(p)$ generated by $\sigma_1|U,\ldots,\sigma_n|U$ and $C(U,\mathbb{K}^n)$ are isomorphic. As the set $\{\sigma_1(y),\ldots,\sigma_n(y)\}$ is a base of E_y, the C(U)-submodule generated by $\sigma_1|U,\ldots,\sigma_n|U$ of $\Gamma_U(p)$ is stalkwise dense. Hence the Stone-Weierstraß theorem (4.3) applied to the bundle $p_{/p^{-1}(U)} : p^{-1}(U) \to U$ shows that the C(U)-submodule generated by $\sigma_1|U,\ldots,\sigma_n|U$ is equal to $\Gamma_U(p)$, i.e. $\Gamma_U(p)$ and $C(U,\mathbb{K}^n)$ are isomorphic. Now apply (14.10) to complete the proof. □

19. The space $\mathrm{Mod}(\Gamma(p), C(X))$.

In this section we shall discuss the existence of $C(X)$-module homomorphisms between the space $\Gamma(p)$ of sections in a bundle $p : E \to X$ and $C_b(X)$. For bundles of Banach spaces it will turn out that this question is closely related with the structure of the closure of the "unit ball" $\{\alpha \in E : \|\alpha\| \leq 1\}$.

Let us start with a full bundle $p : E \to X$ over a completely regular base space X. If $\mathcal{S}$ is any family of precompact subsets of $\Gamma(p)$ whose union generates, then we know from the remarks preceeding (11.21) that there is a bundle $q : F \to X$ such that $\mathrm{Mod}(\Gamma(p), C_b(X))$, equipped with the topology of uniform convergence on elements $S \in \mathcal{S}$, is topologically and algebraically isomorphic to a $C_b(X)$-submodule of $\Gamma(q)$. The stalks of the bundle $q : F \to X$ are subspaces of $E_x' = L(E_x, \mathbb{K})$, where E_x denotes the stalk of $p : E \to X$ over $x \in X$. Note that the choice of the stalks of $q : F \to X$ does not depend on $\mathcal{S}$, although of course the topology on F does.

Let us try to describe the bundle $q : F \to X$ in greater detail. By the remarks in (15.1), we may identify E_x' (the dual space of E_x) with a subspace of $\Gamma(p)'$ and by (11.20), this embedding is topological, if we equip E_x' with the topology of uniform convergence on elements of $\mathcal{S}(x) = \{\varepsilon_x(S) : S \in \mathcal{S}\}$ and $\Gamma(p)'$ with the topology of uniform convergence on $\mathcal{S}$.

Now let $T : \Gamma(p) \to C_b(X)$ be a continuous $C_b(X)$-module homomorphism. Then T corresponds to a section $\lambda_T \in \Gamma(q)$, and T and λ_T are related by the equation

$$T(\sigma)(x) = \lambda_T(x)(\sigma(x)) \qquad \text{for all } x \in X,\ \sigma \in \Gamma(p)$$

Moreover, λ_T may also be viewed as the unique bundle morphism $\lambda_T : E \to X \times \mathbb{K}$ from E into the trivial bundle $pr_1 : X \times \mathbb{K} \to X$ which represents T by (10.7).

Furthermore, as $\lambda_T(x)$ is an element of E_x' for every $x \in X$ and as E_x' may be identified with a subspace of $\Gamma(p)'$, we also may view λ_T as a mapping into $\Gamma(p)'$.

The following result is a generalization of (10.23):

19.1 Proposition. *Let* $p : E \to X$ *be a full bundle over a completely regular base space* X. *Then* $T : \Gamma(p) \to C_b(X)$ *is a continuous* $C_b(X)$*-module homomorphism if and only if there is a uniquely determined* $\sigma(\Gamma(p)',\Gamma(p))$*-continuous mapping* $\lambda_T : X \to \Gamma(p)'$ *satisfying*

(1) $\lambda_T(x) \in E_x'$ for every $x \in X$.

(2) $\lambda_T(X)$ is an equicontinuous subset of $\Gamma(p)'$.

such that $T(\sigma)(x) = \lambda_T(x)(\sigma(x))$ *for all* $x \in X$ *and all* $\sigma \in \Gamma(p)$. *Moreover, if S is a total and directed family of precompact subsets of* $\Gamma(p)$ *and if we equip* $Mod(\Gamma(p),C_b(X))$ *and* $\Gamma(p)'$ *with the topology of uniform convergence on S, then the mapping* $\lambda_T : X \to \Gamma(p)'$ *is continuous and*

$$\begin{array}{rcl} \lambda_- : Mod_S(\Gamma(p),C_b(X)) & \to & C_b(X,\Gamma_S(p)') \\ T & \to & \lambda_T \end{array}$$

is a continuous and injective $C_b(X)$*-module homomorphism which is open onto its image.*

Proof. Let $T : \Gamma(p) \to C_b(X)$ be a continuous $C_b(X)$-module homomorphism and let $\lambda_T : X \to \Gamma(p)'$ be as explained in the above remarks. By construction, we have $\lambda_T(x) \in E_x'$ and $T(\sigma)(x) = \lambda_T(x)(\sigma(x))$ for all $\sigma \in \Gamma(p)$ and all $x \in X$. This last equation also shows the $\sigma(\Gamma(p)',\Gamma(p))$-continuity of λ_T, as for every $\sigma \in \Gamma(p)$ the mapping $x \to \lambda_T(x)(\sigma) = \lambda_T(x)(\sigma(x)) = T(\sigma)(x)$ belongs to $C(X)$. Moreover, the set $\lambda_T(X)$ is equicontinuous, as we have

$$(*)\quad \{\sigma \in \Gamma(p) : |\lambda_T(x)(\sigma)| \le 1 \text{ for all } x \in X\}$$
$$= \{\sigma \in \Gamma(p) : |\lambda_T(x)(\sigma(x))| \le 1 \text{ for all } x \in X\}$$
$$= \{\sigma \in \Gamma(p) : |T(\sigma)(x)| \le 1 \text{ for all } x \in X\}$$
$$= \{\sigma \in \Gamma(p) : \|T(\sigma)\| \le 1\}$$

and the last set is open by the continuity of T.
Conversely, let $\lambda : X \to \Gamma(p)'$ be a $\sigma(\Gamma(p)',\Gamma(p))$-continuous mapping satisfying conditions (1) and (2). Define

$$T_\lambda \;:\; \Gamma(p) \to C_b(X),$$

where $T_\lambda(\sigma)(x) = \lambda(x)(\sigma(x))$ for all $x \in X$. Then $T_\lambda(\sigma) : X \to \mathbb{K}$ is continuous for every $\sigma \in \Gamma(p)$, because we have $T_\lambda(\sigma)(x) = \lambda(x)(\sigma(x)) = \lambda(x)(\sigma)$ and because the mapping $\lambda : X \to \Gamma(p)'$ is $\sigma(\Gamma(p)',\Gamma(p))$-continuous.
Further, the mapping $T_\lambda(\sigma) : X \to \mathbb{K}$ is bounded, since $\lambda(X)$ is equicontinuous and hence weakly bounded.
Using (*) again, we see that the equicontinuity of $\lambda(X)$ implies the continuity of $T_\lambda : \Gamma(p) \to C_b(X)$. Obviously, T_λ is a $C_b(X)$-module homomorphism.

Now let S be a directed and total family of precompact subsets of $\Gamma(p)$ and let $\lambda : X \to \Gamma(p)'$ be any $\sigma(\Gamma(p)',\Gamma(p))$-continuous mapping satisfying (1) and (2). It is an easy consequence of (III.4.5) in [Sch 71] that under these conditions the $\sigma(\Gamma(p)',\Gamma(p))$-topology and the S-topology agree on $\lambda(X)$. Thus, the mapping $\lambda : X \to \Gamma(p)'$ is continuous for the S-topology. It follows that

$$\begin{array}{cccc} \lambda_- & : & \mathrm{Mod}_S(\Gamma(p), C_b(X)) & \to & C_b(X, \Gamma(p)'_S \\ & & T & \to & \lambda_T \end{array}$$

is an injective $C_b(X)$-module homomorphism (note that λ_T is bounded for every $T \in \mathrm{Mod}(\Gamma(p), C_b(X))$ by [Sch 71, III.4.1]).

It remains to show that λ_- is continuous and open onto its image: A typical neighborhood of O in $\mathrm{Mod}_S(\Gamma(p), C_b(X))$ looks like

$$\{T : \sup_{\sigma \in S} \|T(\sigma)\| \leq 1\}$$

for a certain $S \in S$, and a typical neighborhood of O in $C_b(X, \Gamma(p)'_S)$ is given by

$$\{F \in C_b(X, \Gamma(p)'_S : \sup_{x \in X} \sup_{\sigma \in S} |F(x)(\sigma)| \leq 1\}.$$

An easy calculation shows that $\sup_{\sigma \in S} \|T(\sigma)\| \leq 1$ if and only if $\sup_{x \in X} \sup_{\sigma \in S} |\lambda_T(x)(\sigma)| \leq 1$ and the proof is complete. $\square$

19.2 Remarks. (i) Under the conditions of (19.1), we let

$$M_S^p := \{(x, \lambda_T(x)) : T \in \mathrm{Mod}(\Gamma(p), C_b(X)), x \in X\} \subset X \times \Gamma(p)'_S,$$

equipped with the topology induced by the product topology and we let

$$\pi^p : M_S^p \to X$$

be the restriction of the first projection. It follows from (8.4(ii)) that M_S^p is a subbundle of the trivial bundle $X \times \Gamma(p)'_S$. Moreover, an application of (1.6(viii)) yields that the bundle $\pi^p : M_S^p \to X$ and the bundle $q : F \to X$ constructed in section 11 to represent $\mathrm{Mod}_S(\Gamma(p), C_b(X))$ are isomorphic. Let us point out that, in particular, $\mathrm{Mod}_S(\Gamma(p), C_b(X))$ may always be represented as a space of sections in a bundle over X with a Hausdorff bundle space.

(ii) If $\Gamma(p)$ is barreled, especially if $p : E \to X$ is a bundle of Banach spaces, and if the union of S generates $\Gamma(p)$, we may substitute the condition (2) in (19.1) by

(2') $\lambda_T(X)$ is bounded in $\Gamma(p)'$

(see [Sch 71, IV.1.6]).
In this case, $\mathrm{Mod}_S(\Gamma(p), C_b(X))$ is isomorphic to the space of <u>all</u> sections in the bundle $\pi^p : M_S^p \to X$.

We still know very little about the size of the stalks of the bundle $\pi^p : M_S^p \to X$. In fact, there are examples such that all stalks consist of 0 only. Let us desribe some elements of the $\bigcup_{x \in X} E'_x$ which certainly do <u>not</u> belong to M_S^p:

We shall again use the notation introduced in section 16. Especially, if $\nu_j : E \to \mathbb{R}$ is a seminorm of the bundle $p : E \to X$, then $\tilde{\nu}_j : E \to \mathbb{R}$ denotes the largest lower semicontinuous function less than or equal to ν_j. We define the "bad" part of the bundle $p : E \to X$ as follows:

Let F be the intersection of all closed subsets $A \subset E$ such that $A \cap E_x$ is a non-empty linear subspace of E_x for every $x \in X$. Clearly, F contains the closure of $\{0 \in E_x : x \in X\}$. Using the same proof as in (16.4, (ii) $\to$ (iii)), one can show that F is the smallest closed subset of E such that

(i) $E_x \cap F$ is a linear subspace of E_x for every $x \in X$.

(ii) $\tilde{\nu}_j(\alpha) = 0$ for all $j \in J$ implies $\alpha \in F$.

If we define

$$F_x := E_x \cap F \qquad \text{for every } x \in X,$$

then we have:

19.3 Proposition. *Let $p : E \to X$ be a full bundle over a completely regular base space X.*

(i) The stalk over $x \in X$ of the bundle $\pi^p : M_S^p \to X$ is contained in $\{x\}\times F_x^o \subset \{x\}\times E_x'$.

(ii) If the stalks of the bundle $\pi^p : M_S^p \to X$ are all equal to $\{x\}\times E_x'$, $x \in X$, i.e. if for every $x \in X$ and every $\phi \in E_x'$ there is a continuous $C_b(X)$-module homomorphism $T : \Gamma(p) \to C_b(X)$ such that

$$T(\sigma)(x) = \phi(\sigma(x)) \qquad \text{for all } \sigma \in \Gamma(p),$$

then E is a Hausdorff space.

Proof. (i) : Let $x \in X$ and let $(x,\phi) \in (\pi^p)^{-1}(x)$. By construction of the bundle π^p: $M_S^p \to X$ we can find a continuous $C_b(X)$-module homomorphism $T : \Gamma(p) \to C_b(X)$ such that $\lambda_T(x) = \phi$. By the remarks preceeding (19.1), the mapping λ_T may be viewed as a bundle morphism $\lambda_T : E \to X\times \mathbb{K}$ by defining

$$\lambda_T(\alpha) = (p(\alpha), \lambda(p(\alpha))(\alpha)).$$

Let $A = \lambda_T^{-1}(\{(y,0) : y \in X\})$. As λ_T is continuous, the set A is closed. From $A \cap E_y = \lambda_T(y)^{-1}(0)$ we conclude that $A \cap E_y$ is a linear subspace of E_y for every $y \in X$ and hence $F \subset A$. This implies $F \cap E_x = F_x \subsetneq A \cap E_x = \lambda_T(x)^{-1}(0) = \phi^{-1}(0)$, i.e. $\phi \in F_x^o$.

(ii) : If the stalks of the bundle $\pi^p : M_S^p \to X$ are all equal to the E_x', $x \in X$, then F_x^o and E_x' coincide for all $x \in X$ by (i). Using polars, we conclude that $F_x = \{0\}$ and hence $F = \{0 \in E_x : x \in X\}$ is closed in E. Now (16.4) yields that E is Hausdorff. □

Of course, we would like to show that the stalks of the bundle $\pi^p : M_S^p \to X$ are identical with the family $(F_x^o)_{x \in X}$. I do not know an answer to this question at all. However, for a certain type of bundles of Banach spaces, the situation is less hopeless:

19.4 Definition. A bundle $p : E \to X$ is called *separable*, if there is a countable subset $A \subset \Gamma(p)$ such that $\{\sigma(x) : \sigma \in A\}$ is dense in E_x for every $x \in X$. □

19.5 Examples. (i) The trivial bundle $pr_1 : X \times \mathbb{K} \to X$ is always separable; more generally, if E is a separable topological vector space, then $pr_1 : X \times E \to X$ is a separable bundle.

(ii) If E is separable and if X is locally compact and σ-compact, then every locally trivial bundle $p : E \to X$ is separable.

(iii) If $p : E \to X$ is a bundle of finite dimensional vector spaces, if X is a compact metric space and if E is Hausdorff, then the bundle $p : E \to X$ is separable.
(Indeed, let $A_n := \{x \in X : \dim E_x \le n\}$. Then A_n is closed by (18.3) and we have $A_n \subset A_{n+1}$ for all $n \in \mathbb{N}$. As X is metric, we may find a countable family $(B_{n,m})_{m \in \mathbb{N}}$ of closed subsets of X such that

$$\bigcup_{n \in \mathbb{N}} B_{n,m} = A_n \setminus A_{n-1}.$$

From (18.5) we conclude that the bundle $p|p^{-1}(B_{n,m}) : p^{-1}(B_{n,m}) \to B_{n,m}$ is locally trivial. Thus, for every $n \in \mathbb{N}$ and every $m \in \mathbb{N}$ we can find finitely many closed subsets $C_{n,m,j}$, $1 \le j \le k_{n,m}$, such that the bundles $p : p^{-1}(C_{n,m,j}) \to C_{n,m,j}$ are trivial. By (i) we may find countable subsets $A'_{n,m,j} \subset \Gamma_{C_{n,m,j}}(p)$ such that the set $\{\sigma'(x) : \sigma' \in A'_{n,m,j}\}$ is dense in E_x for every $x \in C_{n,m,j}$.
From the Stone-Weierstraß theorem (4.2) we conclude that the restriction map $\sigma \to \sigma|C_{n,m,j} : \Gamma(p) \to \Gamma_{C_{n,m,j}}(p)$ maps $\Gamma(p)$ onto a dense subspace of $\Gamma_{C_{n,m,j}}(p)$. As $\Gamma_{C_{n,m,j}}(p)$ is (topologically) isomorphic to $C_b(C_{n,m,j}, \mathbb{K}^n)$, we conclude that this space is metrizable. Hence we can find a countable subset $A_{n,m,j} \subset \Gamma(p)$ such that the closure of $\{\sigma|C_{n,m,j} : \sigma \in A_{n,m,j}\}$ contains $A'_{n,m,j}$. In particular, we have that the set $\{\sigma(x) : \sigma \in A_{n,m,j}\}$ is dense in E_x for every $x \in C_{n,m,j}$.
Finally, we set

$$A = \bigcup_{n \in \mathbb{N}} \bigcup_{m \in \mathbb{N}} \bigcup_{1 \le j \le k_{n,m}} A_{n,m,j}.)$$

For more examples and results concerning separable bundles, we refer to the papers of M. Dupré (see for example [Du 73]). The paper just mentioned contains also the idea of the proof of (ii) as it was given above. Note however that M.Dupré uses a more special type of bundles.

From now on we shall equip the spaces $\Gamma(p)'$ and $\mathrm{Mod}(\Gamma(p), C_b(X))$ always with the topology of pointwise convergence and we shall denote these spaces by $\Gamma_s(p)'$ and $\mathrm{Mod}_s(\Gamma(p), C_b(X))$, resp.. Moreover, we shall again use the notations of section 15, which we shall recall for convenience:

$$B_1^o \; : \; \text{unit ball of } \Gamma(p)' ,$$

$$\mathcal{B}_x \ : \ \text{unit ball of } E'_x \subset \Gamma(p)' \ ,$$

$$\mathcal{B}_A \ = \bigcup_{x \in A} \mathcal{B}_x \ .$$

19.6 Proposition. *Let* $p : E \to X$ *be a bundle of Banach spaces over a compact base space* X*. Then the bundle* $p : E \to X$ *is separable if and only if* 0 *has a countable neighborhood base in* $\mathcal{B}_X$.

Proof. If $p : E \to X$ is separable, then choose a countable subset $A \subset \Gamma(p)$ such that $\{\sigma(x) : \sigma \in A\}$ is dense in E_x for every $x \in X$. For every $\sigma \in A$ we let

$$A_\sigma \ := \ \{\phi \in \mathcal{B}_X : |\phi(\sigma)| \leq 1\}.$$

Then A_σ is a closed neighborhood of 0 in $\mathcal{B}_X$. Moreover, if $\phi \in E'_x$ is given, then

$$\begin{aligned} \phi \in \bigcap_{\sigma \in A} A_\sigma \quad & \text{if and only if} \quad |\phi(\sigma)| \leq 1 \text{ for all } \sigma \in A \\ & \text{if and only if} \quad |\phi(\sigma(x))| \leq 1 \text{ for all } \sigma \in A \\ & \text{if and only if} \quad |\phi(\alpha)| \leq 1 \text{ for all } \alpha \in E_x \\ & \text{if and only if} \quad \phi = 0. \end{aligned}$$

We conclude that

$$\bigcap_{\sigma \in A} A_\sigma \ = \ \{0\}.$$

Since $(A_\sigma)_{\sigma \in A}$ is countable and since $\mathcal{B}_X$ is compact, we conclude that 0 has a countable neighborhood base in $\mathcal{B}_X$.

Conversely, assume that 0 has a countable base $(U_n)_{n \in \mathbb{N}}$. Then for each $n \in \mathbb{N}$ we may pick elements $\sigma_{n,1},\ldots,\sigma_{n,m_n} \in \Gamma(p)$ such that

$$\{\phi \in \mathcal{B}_X : |\phi(\sigma_{n,j})| \leq 1 \text{ for } 1 \leq j \leq m_n\} \subset U_n.$$

Let $A \subset \Gamma(p)$ be the linear subspace over the rational numbers generated by $\{\sigma_{n,j} : n \in \mathbb{N} \text{ and } 1 \leq j \leq m_n\}$. Clearly, the set A is

countable. We wish to show that $\{\sigma(x) : \sigma \in A\} =: A_x$ is dense in E_x for every $x \in X$. As the closure of A_x coincides with A_x^{oo}, we have to show that $\phi = 0$ whenever ϕ belongs to E_x', has norm less than or equal to 1 and satisfies $\phi(A_x) = \{0\}$.
If $\phi(A_x) = \{0\}$, then $|\phi(\sigma_{n,j})(x)| = |\phi(\sigma_{n,j})| = 0$ for all $n \in \mathbb{N}$ and all $1 \leq j \leq m_n$. We conclude that $\phi \in U_n$ for every $n \in \mathbb{N}$ and therefore $\phi \in \bigcap_{n \in \mathbb{N}} U_n = \{0\}$. □

We are now ready for the construction of $C(X)$-module homomorphisms $T \in \mathrm{Mod}(\Gamma(p), C(X))$. This construction will use to a large extent the ideas of Douady and dal Soglio-Hérault as they were presented in section 3.

19.7 Proposition. *Let $p : E \to X$ be a separable bundle of Banach spaces over a compact base space X and let $K \subset B_X$ be a subset such that*

(i) $0 \in K$

(ii) $B_x \cap K$ is closed, convex and symmetric for all $x \in X$

(iii) If $U \subset \Gamma_s(p)'$ is open, then the set $\{x \in X : B_x \cap K \cap U \neq \emptyset\}$ is open in X.

Then for every $x_0 \in X$ and every $\phi_0 \in K \cap B_{x_0}$ there is a continuous function $\eta : X \to \Gamma_s(p)'$ such that $\eta(x_0) = \phi_0$ and $\eta(x) \in B_x \cap K \subset E_x'$ for every $x \in X$.

Proof. From (19.6) we know that 0 has a countable neighborhood base $(U_n)_{n \in \mathbb{N}}$ in B_X. We may assume that the U_n, $n \in \mathbb{N}$, have the following property:

There exists a sequence $(V_n)_{n \in \mathbb{N}}$ of open subsets of $\Gamma_s(p)'$ such that:

(1) $U_n = \mathcal{B}_X \cap V_n$ for each $n \in \mathbb{N}$.

(2) V_n is a convex, symmetric neighborhood of 0.

(3) $\overline{V}_{n+1} + \overline{V}_{n+1} \subset V_n$

A *selection* $\eta : X \to \mathcal{B}_X$ is a mapping such that $\eta(x) \in \mathcal{B}_X$ for every $x \in X$. Let V be an open neighborhood of 0 in $\Gamma_s(p)'$. We say that a selection $\eta : X \to \mathcal{B}_X$ is V-*continuous*, if every $x \in X$ has an open neighborhood W such that $\eta(y) - \eta(x) \in V$ for all $y \in W$.

In the following we abbreviate: $K_x = K \cap \mathcal{B}_x$.

We shall divide the proof of (19.7) into a series of lemmas:

19.8 Lemma. *Under the assumptions of* (19.7) *we have*

(i) *If a selection* $\eta : X \to \mathcal{B}_X$ *is* V_n*-continuous for every* $n \in \mathbb{N}$, *then* η *is a continuous mapping.*

(ii) *If* $(\eta_m)_{m \in \mathbb{N}}$ *is a sequence of* V_{n+2}*-continuous selections such that*

(CF) For every $k \in \mathbb{N}$ there is an $N \in \mathbb{N}$ such that for all pairs $m_1, m_2 \in \mathbb{N}$ with $m_1, m_2 \geq N$ and all $x \in X$ we have $\eta_{m_1}(x) - \eta_{m_2}(x) \in V_k$.

holds, then $\lim_{m\to\infty} \eta_m$ *defined by* $(\lim_{m\to\infty} \eta_m)(x) = \lim_{m\to\infty} \eta_m(x)$ *exists and is* V_n*-continuous.*

(iii) *Let* $(\eta_n)_{n \in \mathbb{N}}$ *be a sequence of selections such that*

(1) η_n is V_n-continuous

(2) $\eta_{n+1}(x) - \eta_n(x) \in V_{n-1}$ for all $n \in \mathbb{N}$.

then $\lim_{n\to\infty} \eta_n$ *exists and is continuous.*

Proof. (i): If $\eta(x) = 0$ and if V is any neighborhood of 0, then there is an $n \in \mathbb{N}$ so that $\mathcal{B}_X \cap V_n \subset V$. As η is V_n-continuous, we can find a neighborhood W of x such that $\eta(y) - \eta(x) = \eta(y) - 0 = \eta(y) \in V_n$ for all $y \in W$. Hence the selection η is continuous at x.

Now assume that $\eta(x) \neq 0$. Let W be an open neighborhood of x. Then $\mathcal{B}_{X\setminus W}$ is closed and does not contain $\eta(x)$. Therefore the set $\mathcal{B}_X \setminus \mathcal{B}_{X\setminus W} = \mathcal{B}_W \setminus \{0\}$ is an open neighborhood of $\eta(x)$. We conclude that $\mathcal{B}_{\overline{W}}$ is a closed neighborhood of $\eta(x)$ and that

$$\cap \{\mathcal{B}_{\overline{W}} : x \in W^o\} = \mathcal{B}_x.$$

Note that this intersection is directed by inclusion.
Moreover, for an element $\phi \in \Gamma_s(p)'$ we have

$$\phi \in \bigcap_n (\eta(x) + \overline{V}_n) \cap \mathcal{B}_x \text{ iff } \phi \in \mathcal{B}_x \text{ and } \phi - \eta(x) \in \overline{V}_n \text{ for all } n$$
$$\text{iff } \phi = \eta(x).$$

Hence the family of sets

$$\{\mathcal{B}_{\overline{W}} \cap (\eta(x) + \overline{V}_n) : x \in W^o,\ n \in \mathbb{N}\}$$

is a filtered system of closed neighborhoods in $\mathcal{B}_x$ of $\eta(x)$ having intersection $\{\eta(x)\}$, i.e. it is a neighborhood base of $\eta(x)$ in $\mathcal{B}_x$.

Now let V be any open neighborhood of $\eta(x)$. Then there is an open neighborhood W_1 of x and a natural number $n \in \mathbb{N}$ such that

$$\mathcal{B}_{W_1} \cap (\eta(x) + V_n) \subset V.$$

As η is V_n-continuous, we can find a neighborhood $W \subset W_1$ of x such that $\eta(y) - \eta(x) \in V_n$ for all $y \in W$. We conclude that

$$\begin{aligned}\eta(W) &\subset (\eta(x) + V_n) \cap \mathcal{B}_W \\ &\subset (\eta(x) + V_n) \cap \mathcal{B}_{W_1} \\ &\subset V,\end{aligned}$$

i.e. η is continuous at x.

(ii) : The family $(V_n \cap \mathcal{B}_x)_{n\in\mathbb{N}}$ is a neighborhood base at 0 in $\mathcal{B}_x$. Hence, by assumption, the sequence $(\eta_m(x))_{m\in\mathbb{N}}$ is a Cauchy sequence in $\mathcal{B}_x$. As $\mathcal{B}_x$ is complete (being compact) $\lim_{m\to\infty} \eta_m(x)$ exists. Define

$$\eta(x) \quad := \quad \lim_{m\to\infty} \eta_m(x).$$

Then η is a selection, which is V_n-continuous:
Indeed, let $x \in X$. Then there is a natural number $N \in \mathbb{N}$ such that $\eta_{m_1}(x) - \eta_{m_2}(x) \in V_{n+3}$ for all $m_1, m_2 \geq N$ and all $x \in X$. For every $x \in X$, this implies

$$\begin{aligned} \eta_N(x) - \eta(x) \quad &= \quad \eta_N(x) - \lim_{m\to\infty} \eta_m(x) \\ &\in \quad \overline{V}_{n+3} \\ &\subset \quad V_{n+2}. \end{aligned}$$

Now let $x_o \in X$. As η_N is V_{n+2}-continuous, there is an open neighborhood W of x_o such that $\eta_N(y) - \eta_N(x_o) \in V_{n+2}$ for all $y \in W$. For a given $y \in W$ this implies:

$$\begin{aligned} \eta(y) - \eta(x_o) \quad &= \quad \eta(y) - \eta_N(y) + \eta_N(y) - \eta_N(x_o) + \eta_N(x_o) - \eta(x) \\ &\in \quad V_{n+2} \qquad + \qquad V_{n+2} \qquad + \qquad V_{n+2} \\ &\subset \quad V_n. \end{aligned}$$

(iii) : Firstly, note that for $m \geq n$ we have

$$\begin{aligned} \eta_m(x) - \eta_n(x) \quad &= \quad \eta_m(x) - \eta_{m-1}(x) + \eta_{m-1}(x) - \ldots + \eta_{n+1}(x) - \eta_n(x) \\ &\in \quad (V_{m-2} + V_{m-3}) + V_{m-4} + V_{m-5} + \ldots + V_n + V_{n-1} \\ &\subset \quad (V_{m-4} + V_{m-4}) + V_{m-5} + \ldots + V_n + V_{n-1} \\ &\subset \quad (V_{m-5} + V_{m-5}) + \ldots + V_n + V_{n-1} \\ &\subset \quad \ldots \\ &\subset \quad V_{n-1} + V_{n-1} \\ &\subset \quad V_{n-2}. \end{aligned}$$

Hence (ii) shows that $\lim_{n\to\infty} \eta_n = \eta$ exists and that η is V_n-continuous for every $n \in \mathbb{N}$. Thus, the selection η is continuous by (i). □

19.9 Lemma. *If* $f_i : X \to [0,1]$, $1 \le i \le n$, *are continuous functions such that* $\sum_{i=1}^{n} f_i = 1$ *and if* $\eta_i : X \to \mathcal{B}_X$, $1 \le i \le n$, *are* V_{m+1}*-continuous, then the selection* $\sum_{i=1}^{n} f_i \cdot \eta_i$ *is* V_m*-continuous.*

Proof. Firstly, note that

$$\bigcap_{r>0} r \cdot \mathcal{B}_X = \{0\}.$$

Hence there is a real number $r > 0$ such that $r \cdot \mathcal{B}_X \subset V_{n+m+1}$. Fix $x_o \in X$. Then there is an open neighborhood W of x such that

$$(1) \qquad \eta_i(y) - \eta_i(x_o) \in V_{m+1} \quad \text{for } 1 \le i \le n \text{ and } y \in W$$

$$(2) \qquad |f_i(y) - f_i(x_o)| < r \quad \text{for all } 1 \le i \le n \text{ and all } y \in W.$$

Let $y \in W$ be arbitrary. Then we have

$$\Big(\sum_{i=1}^{n} f_i \cdot \eta_i\Big)(y) - \Big(\sum_{i=1}^{n} f_i \cdot \eta_i\Big)(x_o) =$$

$$= \sum_{i=1}^{n} (f_i(y) \cdot \eta_i(y) - f_i(x_o) \cdot \eta_i(x_o))$$

$$= \sum_{i=1}^{n} (f_i(y) \cdot \eta_i(y) - f_i(x_o) \cdot \eta_i(y) + f_i(x_o) \cdot \eta_i(y) - f_i(x_o) \cdot \eta_i(x_o))$$

$$= \sum_{i=1}^{n} (f_i(y) - f_i(x_o)) \cdot \eta_i(y) + \sum_{i=1}^{n} f_i(x_o) \cdot (\eta_i(y) - \eta_i(x_o)).$$

As the $f_i(x_o)$ sum up to 1, as $\eta_i(y) - \eta_i(x_o)$ belongs always to V_{m+1} and as V_{m+1} is convex, we obtain

$$\sum_{i=1}^{n} f_i(x_o) \cdot (\eta_i(y) - \eta_i(x_o)) \in V_{m+1}$$

As $|f_i(y) - f_i(x_o)| < r$, we have $(f_i(y) - f_i(x_o)) \cdot \eta_i(y) \in V_{m+n+1}$, i.e.

$$\sum_{i=1}^{n} (f_i(y) - f_i(x_o)) \cdot \eta_i(y) \in V_{m+1}$$

Together, this yields

$$\left(\sum_{i=1}^{n} f_i\cdot\eta_i\right)(y) - \left(\sum_{i=1}^{n} f_i\cdot\eta_i\right)(x_o) \in V_{m+1} + V_{m+1} \subset V_m ,$$

i.e. $\sum_{i=1}^{n} f_i\cdot\eta_i$ is V_m-continuous. □

19.10 *Let* $\phi \in K_{x_o}$. *Then for every* $n \in \mathbb{N}$ *there is a* V_n*-continuous selection* η *such that*

(a) $\eta(x) \in K_x$ *for all* $x \in X$.

(b) $\eta(x_o) = \phi$.

Proof. Let

$$U = \{x \in X : K_x \cap (\phi + V_{n+3}) \neq \emptyset\}.$$

Then U is an open neighborhood of x_o. Moreover, we define $\phi_x = 0$ if $x \notin U$ and $\phi_{x_o} := \phi$. If $x_o \neq x \in U$ is given, then let ϕ_x be an arbitrary element of $K_x \cap (\phi + V_{n+3})$.
Now fix a continuous function $f : X \to [0,1]$ such that f vanishes on $X \setminus U$ and takes the value 1 at x_o. Define

$$\eta : X \to \bigcup_{x\in X} K_x$$

by

$$\eta(x) = f(x)\cdot\eta_x.$$

Then we have $\eta(x_o) = \phi$. Furthermore, the selection η is V_n-continuous: Indeed, let $y_o \in U$. As in the proof of (19.9) we choose a real number $r > 0$ such that $r\cdot B_X \subset V_{n+3}$. Let $V \subset U$ be any neighborhood of y_o such that $|f(y) - f(y_o)| < r$ for all $y \in V$. Then for a given $y \in V$ we have

$$\begin{aligned}\eta(y) - \eta(y_o) &= f(y)\cdot\eta_y - f(y_o)\cdot\eta_{y_o}\\ &= (f(y) - f(y_o))\cdot\eta_y + f(y_o)\cdot(\eta_y - \eta_{y_o})\\ &\in V_{n+3} + V_{n+3} \subset V_{n+2} \subset V_n\end{aligned}$$

If $y_o \notin U$, then $f(y_o) = 0 < r$. Hence we may find an open neighborhood U' of y_o such that $|f(y)| < r$ for all $y \in U'$. Thus, for every $y \in U'$ we have

$$\begin{aligned}\eta(y) - \eta(y_o) &= f(y)\cdot\eta_y - f(y_o)\cdot\eta_{y_o} \\ &= f(y)\cdot\eta_y \\ &\in V_{n+3} \subset V_n. \quad \square\end{aligned}$$

19.11 Lemma. *Let* $n \in \mathbb{N}$ *be a natural number, let* $x_o \in X$ *and let* $\eta : X \to K$ *be a* V_n*-continuous selection. Then there is a* V_{n+1}*-continuous selection* $\eta' : X \to K$ *such that*

(i) $\eta(x_o) = \eta'(x_o)$

(ii) $\eta(x) - \eta'(x) \in V_{n-1}$ *for all* $x \in X$.

Proof. Let $x \in X$ be fixed for a moment. By (19.10) there is a V_{n+2}-continuous selection η_x such that $\eta_x(x) = \eta(x)$. As η is V_n-continuous and as η_x is V_{n+2}-continuous, there is an open set U_x around x such that for every $y \in U_x$ we have $\eta(y) - \eta(x) \in V_n$ and $\eta_x(y) - \eta_x(x) \in V_{n+2}$. We may assume that x_o does not belong to U_x provided that $x \neq x_o$. Moreover, for every $y \in U_x$ we have

$$\begin{aligned}\eta(y) - \eta_x(y) &= \eta(y) - \eta(x) + \eta(x) - \eta_x(y) \\ &= \eta(y) - \eta(x) + \eta_x(x) - \eta_x(y) \\ &\in V_n + V_{n+2}.\end{aligned}$$

Now the U_x cover the compact space X; hence there are finitely many elements $x_1,\dots,x_n \in X$ such that $U_{x_1} \cup \dots \cup U_{x_n} = X$. Since x_o occurs in exactly one of the U_x, this element belongs to $\{x_1,\dots,x_n\}$; w.l.o.g. we may assume that $x_o = x_1$. Let $(f_i)_{i=1}^n$ be a partition of unity subordinate to the covering $(U_{x_i})_{i=1}^n$. Then $f_1(x_o) = 1$ and $f_i(x_o) = 0$ for $2 \le i \le n$. We define

$$\eta' := \sum_{i=1}^n f_i\cdot\eta_{x_i}.$$

Then $\eta'(x_o) = \eta(x_o)$. Moreover, the selection η' is V_{n+1}-continuous by (19.9). Finally, for a given $x \in X$ let

$$J_x := \{i \in \{1,\ldots,n\} : x \in U_{x_i}\}.$$

Since $x \notin U_{x_i}$ implies $f_i(x) = 0$, we have

$$\sum_{i \in J_x} f_i(x) = 1.$$

Moreover, as $x \in U_{x_i}$ implies $\eta(x) - \eta_{x_i}(x) \in V_n + V_{n+2}$ and since $V_n + V_{n+2}$ is convex, we conclude

$$\begin{aligned}
\eta(x) - \eta'(x) &= \eta(x) - \sum_{i=1}^{n} f_i(x)\cdot\eta_{x_i}(x) \\
&= \sum_{i=1}^{n} f_i(x)\cdot(\eta(x) - \eta_{x_i}(x)) \\
&= \sum_{i \in J_x} f_i(x)\ (\eta(x) - \eta_{x_i}(x)) \\
&\in V_n + V_{n+2} \\
&\subset V_n + V_n \\
&\subset V_{n-1}. \quad \square
\end{aligned}$$

We now finish the proof of (19.7):

Firstly, by induction using (19.10) and (19.11), we find a sequence of selections $(\eta_n)_{n \in \mathbb{N}}$ such that

(1) η_n is V_n-continuous.

(2) $\eta_{n+1}(x) - \eta_n(x) \in V_{n-1}$ for all $x \in X$.

(3) $\eta_n(x_o) = \phi_o$ for all $n \in \mathbb{N}$.

(4) $\eta_n(x) \in K_x$ for all $n \in \mathbb{N}$ and all $x \in X$.

Now let $\eta := \lim_{n\to\infty} \eta_n$. Then η exists and is continuous by (19.8). Moreover, $\eta(x_o) = \phi_o$ and $\eta(x) \in K_x$ for all $x \in X$, as K_x is always closed. $\square$

The following proposition states the converse of (19.7):

19.12 Proposition. *Let* $p : E \to X$ *be a bundle of Banach spaces with a compact base space* X. *Define*

$$\tilde{K}_x := \{\phi \in B_x : \phi = \eta(x) \text{ for some continuous selection } \eta : X \to B_X\}$$

Then $\tilde{K}_x$ *is convex and circled. If* $U \subset \Gamma_s(p)'$ *is open, then so is* $\{x \in X : \tilde{K}_x \cap U \neq \emptyset\}$. *Moreover, if* $p : E \to X$ *is a separable bundle, then every* $\tilde{K}_x$ *is closed.*

Proof. Obviously, the set $\tilde{K}_x$ is convex and circled. To show the closedness of $\tilde{K}_x$ for separable bundles $p : E \to X$, let $U_n \subset B_X$ and $V_n \subset \Gamma_s(p)'$, $n \in \mathbb{N}$, be as in the proof of (19.7). Moreover, let ϕ belong to the closure of $\tilde{K}_x$. As B_x is metric, there is a sequence of elements $\phi_n \in \tilde{K}_x$ such that $\lim\limits_{n\to\infty} \phi_n = \phi$. Picking an appropriate subsequence, we may assume that $\phi_{n+1} - \phi_n \in V_{n+1}$ for all $n \in \mathbb{N}$. We define recursively a sequence of continuous selections $\eta_n : X \to B_X$ such that

(1) $\eta_n(x) = \phi_n$

(2) $\eta_{n+1}(y) - \eta_n(y) \in V_{n+1}$ for all $n \in \mathbb{N}$, $y \in X$.

Choose any η_1 satisfying (1).

If η_n is already defined, choose any continuous selection $\xi_{n+1} : X \to B_X$ such that $\xi_{n+1}(x) = \phi_{n+1}$. Then we conclude that $\xi_{n+1}(x) - \eta_n(x) \in V_{n+1}$. As ξ_{n+1} and η_n are continuous, there is an open neighborhood W of x such that $\xi_{n+1}(y) - \eta_n(y) \in V_{n+1}$ for all $y \in W$. Pick a continuous function $f : X \to [0,1]$ such that $f(x) = 1$ and $f(X \setminus W) = \{0\}$. We now define

$$\eta_{n+1} = f\cdot\xi_{n+1} + (1 - f)\cdot\eta_n.$$

Then we compute that $\eta_{n+1}(x) = \xi_{n+1}(x) = \phi_{n+1}$. Moreover, the continuous mapping $\eta_{n+1} : X \to \mathcal{B}_X$ is a selection, as $\mathcal{B}_X$ is always convex.

The fact that V_{n+1} is convex and contains 0 implies that $(\eta_{n+1} - \eta_n)(y) = f(y)\cdot(\xi_{n+1}(y) - \eta_n(y)) \in V_{n+1}$ for all $y \in X$, as f vanishes on $X \setminus W$.

From (19.8(iii)) we conclude that the function $\eta : X \to \mathcal{B}_X$ defined by $\eta(y) = \lim_{n\to\infty} \eta_n(y)$ is a continuous selection. Obviously, we have $\eta(x) = \lim_{n\to\infty} \eta_n(x) = \lim_{n\to\infty} \phi_n = \phi$, i.e. $\tilde{K}_x$ is closed.

Finally, let $U \subset \Gamma_s(p)'$ be open and assume that $\tilde{K}_{x_o} \cap U \neq \emptyset$. Then there is a continuous selection $\eta : X \to \mathcal{B}_X$ such that $\eta(x_o) \in \tilde{K}_{x_o} \cap U$. As η is continuous, $\eta^{-1}(U)$ is an open neighborhood of x_o and by definition of the $\tilde{K}_x$, $x \in X$, we have $\eta(x) \in \tilde{K}_x \cap U$ whenever $x \in \eta^{-1}(U)$. This establishes the fact that $\{x \in X : \tilde{K}_x \cap U \neq \emptyset\}$ is open. □

19.13 Theorem. *Let* $p : E \to X$ *be a separable bundle of Banach spaces over a compact base space* X. *Moreover, let* C *be the smallest closed subset of* E *containing the "unit ball"* $\{\alpha \in E : \|\alpha\| \leq 1\}$ *and having the property that* $C \cap E_x$ *is convex for every* $x \in X$ *. Finally, let* $K_x = (C \cap E_x)^o \subset \mathcal{B}_x$.
Then for every $x \in X$ *and every* $\phi \in K_x$ *there is a continuous* C(X)*-module homomorphism* $T \in \mathrm{Mod}(\Gamma(p),C(X))$ *with* $\|T\| \leq 1$ *and* $\lambda_T(x) = \phi$.
Conversely, if $T \in \mathrm{Mod}(\Gamma(p),C(X))$ *and if* $\|T\| \leq 1$, *then* $\lambda_T(x) \in K_x$ *for every* $x \in X$.

Proof. By (16.9) and (16.10), the set $K := \bigcup_{x\in X} K_x$ satisfies the properties (i)-(iii) of (19.7). Hence, given a point $x \in X$ and an element $\phi \in K_x$, there is a continuous selection $\lambda : X \to K$ such that

$\lambda(x) = \phi$. By (19.1) and (19.2(ii)) the mapping $T_\lambda : \Gamma(p) \to C(X)$ is a continuous $C(X)$-module homomorphism and we have $\lambda_{T_\lambda}(x) = \lambda(x) = \phi$ Viewing λ as a bundle morphism from from E into $X \times \mathbb{K}$, we obtain from (10.13) the equation

$$\|T_\lambda\| = \sup \{ \|\lambda(y)\| : y \in X\}.$$

As $\lambda(y)$ belongs to B_y for every $y \in X$, we have always $\|\lambda(y)\| \leq 1$ and therefore $\|T\| \leq 1$.

Conversely, again by (16.9) and (16.10) the family $(K_x)_{x \in X}$ is the largest family such that $\{x \in X : K_x \cap U \neq \emptyset\}$ is open whenever $U \subset \Gamma_s(p)'$ is open. Thus, by (19.12) the set $\{\phi \in B_x : \phi = \eta(x)$ for some continuous selection $\eta : X \to B_X\}$ is contained in K_x. Now let $T \in Mod(\Gamma(p), C(X))$ and suppose that $\|T\| \leq 1$. From (10.13) we conclude that under these conditions we have for every $x \in X$ the inequality $\|\lambda_T(x)\| \leq 1$. Hence the mapping $\lambda_T : X \to \Gamma(p)'$ maps X into B_X and therefore is a continuous selection by (19.1), i.e. $\lambda_T(x) \in K_x$ for every $x \in X$. □

As we always have the relation $K_x^o = (C \cap E_x)^{oo} = C \cap E_x$, we obtain as a corollary:

19.14 Corollary. *Let* $p : E \to X$ *be a separable bundle of Banach spaces over a compact base space* X. *Then the set*

$\{\alpha \in E: |\lambda(\alpha)| \leq 1$ for all bundle morphisms $\lambda : E \to X \times \mathbb{K}$ with $\|\lambda\| \leq 1\}$

$= \{\alpha \in E: |T(\sigma)(p(\alpha))| \leq 1$ for all $\sigma \in \Gamma(p)$ with $\sigma(p(\alpha)) = \alpha$ and all $T \in Mod(\Gamma(p), C(X))$ with $\|T\| \leq 1\}$

is the smallest closed subset of E *which is stalkwise convex and contains* $\{\alpha \in E : \|\alpha\| \leq 1\}$. □

For every $x \in X$ let G_x be the largest closed vector subspace of $C \cap E_x$ (this space exists as $C \cap E_x$ is convex and closed). From duality between E_x and E'_x we conclude that the vector space generated by K_x is dense in G^o_x. Hence we have:

19.15 Corollary. *Let* $p : E \to X$ *be a separable bundle of Banach spaces over a compact base space* X. *Then the stalks of the bundle* $\pi^p : M^p_S \to X$ *defined in* (19.2) *are dense subspaces of* G^o_x. □

In our next corollary, we characterize bundles with continuous norms via some properties of $\mathrm{Mod}(\Gamma(p), C(X))$:

19.16 Corollary. *Let* $p : E \to X$ *be a separable bundle of Banach spaces and assume that the base space* X *is compact. Then the following conditions are equivalent:*

(i) $p : E \to X$ has continuous norm.

(ii) If $x \in X$ and if $\phi \in E'_x$, then there is a continuous $C(X)$-module homomorphism $T \in \mathrm{Mod}(\Gamma(p), C(X))$ such that $\lambda_T(x) = \phi$ and $||T|| = ||\phi||$.

(iii) If $\sigma \in \Gamma(p)$, then
$\mathrm{norm}(\sigma) = \sup \{ |T(\sigma)| : ||T|| \le 1, T \in \mathrm{Mod}(\Gamma(p), C(X))\}$,
where the mapping $\mathrm{norm}(\sigma) : X \to \mathbb{R}$ is defined by $\mathrm{norm}(\sigma)(x) = ||\sigma(x)||$, where $|T(\sigma)|$ is defined by $|T(\sigma)|(x) = |T(\sigma)(x)|$ in $\mathbb{K}$.

(iv) $||\sigma|| = \sup \{ ||T(\sigma)|| : T \in \mathrm{Mod}(\Gamma(p), C(X))$ and $||T|| \le 1\}$
and the set
$K := \{\phi \in B_X : \phi = \lambda_T(x)$ for some $x \in X$ and some $T \in \mathrm{Mod}(\Gamma(p), C(X))$ with $||T|| \le 1\}$
is closed.

Moreover, under these conditions, the stalks of the bundle

$\pi^p : M_S^p \to X$ *defined in* (19.2) *are equal to* E'_x, $x \in X$, *and the space* $\mathrm{Mod}(\Gamma(p),C(X))$ *separates the points of* $\Gamma(p)$.

Proof. If $p : E \to X$ has continuous norm, then $\{\alpha \in E : \|\alpha\| \le 1\}$ is closed and stalkwise convex; thus (i) implies (ii) by (19.13). Obviously, (ii) implies (iii).

(iii) → (i): By (iii), the mapping $x \to \|\sigma(x)\|$ is a pointwise supremum of continuous functions for every $\sigma \in \Gamma(p)$ and therefore lower semicontinuous. As it is always upper semicontinuous, we have shown (i).

As (iii) implies $\|\sigma\| = \sup\{\|T(\sigma)\| : \|T\| \le 1$ and $T \in \mathrm{Mod}(\Gamma(p), C(X))\}$ and as (ii) shows that $K = \mathcal{B}_X$, it remains to establish the implication (iv) → (ii), i.e. we have to show that under the presence of (iv) the equation $K = \mathcal{B}_X$ holds.

Applying (iv), we obtain:

$$\begin{aligned}
\overline{\mathrm{conv}}\, K &= K^{oo} \\
&= \{\sigma \in \Gamma(p) : |\phi(\sigma)| \le 1 \text{ for all } \phi \in K\}^o \\
&= \{\sigma \in \Gamma(p) : |T(\sigma)(x)| \le 1 \text{ for all } x \in X \text{ and all } \\
&\qquad T \in \mathrm{Mod}(\Gamma(p),C(X)) \text{ with } \|T\| \le 1\}^o \\
&= \{\sigma \in \Gamma(p) : \|T(\sigma)\| \le 1 \text{ for all } T \in \mathrm{Mod}(\Gamma(p),C(X)) \text{ with } \\
&\qquad \|T\| \le 1\}^o \\
&= \{\sigma \in \Gamma(p) : \|\sigma\| \le 1\}^o \\
&= B_1^o.
\end{aligned}$$

As $K \cap \mathcal{B}_X$ is closed and convex, we conclude from (15.15(i)) that $K \cap \mathcal{B}_X = \overline{\mathrm{conv}}(K \cap \mathcal{B}_X) = \overline{\mathrm{conv}}(K) \cap \mathcal{B}_X = B_1^o \cap \mathcal{B}_X = \mathcal{B}_X$ and thus $K = \mathcal{B}_X$. □

The bundle constructed in example (16.3) shows that K is not closed in general, even if we postulate in addition that $\|\sigma\| = \sup\{\|T(\sigma)\| : T \in \mathrm{Mod}(\Gamma(p),C(X)),\ \|T\| \le 1\}$.

Let us conclude this section with a couple of open problems:

1.) Suppose that E is Hausdorff and assume that all stalks of the bundle $p : E \to X$ are finite dimensional. Given a point $x \in X$ and $\phi \in E'_x$, is there a continuous $C(X)$-module homomorphism $T \in \mathrm{Mod}(\Gamma(p), C(X))$ such that $\lambda_T(x) = \phi$?

2.) Let us assume that the set K as it was defined in (19.16(iv)) is closed and suppose that for every $\phi \in E'_x$ there is a $T \in \mathrm{Mod}(\Gamma(p), C(X))$ such that $\lambda_T(x) = \phi$. Define a norm $|||\cdot|||$ on $\Gamma(p)$ by

$$|||\sigma||| = \sup \{ ||T(\sigma)|| : T \in \mathrm{Mod}(\Gamma(p), C(X)), \; ||T|| \leq 1 \}.$$

It is possible to show that $\Gamma(p)$ is a locally $C(X)$-convex $C(X)$-module in this new norm. Moreover, the bundle $p' : E' \to X$ representing $\Gamma(p)$ in this new norm has up to isomorphy (not isometry) the same stalks as the bundle $p : E \to X$ and, in addition, continuous norm.

Question: Is the norm $|||\cdot|||$ equivalent to the orginial norm on $\Gamma(p)$? Is that true in the case where all stalks are finite dimensional?

20. Internal duality of C(X)-modules.

Let E be a Banach space and let E_c' be the dual of E equipped with the topology of compact convergence. From the Mackey-Arens theorem we know that E is (topologically) isomorphic to $(E_c')_c'$. In this section we shall study to what extend these results remain true for the space $\Gamma(p)$ of all sections in a bundle and its "internal dual" $\mathrm{Mod}_c(\Gamma(p),C(X))$. It will turn out that locally trivial bundles are "internal Mackey spaces" in this sense and that for certain bundles with continuous norm, the space $\Gamma(p)$ is at least algebraically isomorphic to its bi-dual $\mathrm{Mod}(\mathrm{Mod}_c(\Gamma(p),C(X)),C(X))$.

20.1 Proposition. *Let* $p : E \to X$ *be a bundle and let* S *be any directed family of bounded subsets of* $\Gamma(p)$ *whose union generates* $\Gamma(p)$. *If* $\sigma \in \Gamma(p)$ *is a section, then the mapping*

$$\tilde{\sigma} : \mathrm{Mod}_S(\Gamma(p),C_b(X)) \to C_b(X)$$
$$T \to T(\sigma)$$

is a continuous $C_b(X)$*-module homomorphism.* □

20.2 Proposition. *Let* $p : E \to X$ *be a full bundle over a completely regular base space* X *and let* S *be a directed family of compact, convex and circled subsets of* $\Gamma(p)$ *whose union generates* $\Gamma(p)$. *If* $\Phi : \mathrm{Mod}_S(\Gamma(p),C_b(X)) \to C_b(X)$ *is a continuous* $C_b(X)$*-module homomorphism, then there exists a (not necessarily continuous) selection* $s : X \to E$ *such that* $\Phi(T)(x) = \lambda_T(x)(s(x))$ *for all* $x \in X$ *and all* $T \in \mathrm{Mod}(\Gamma(p),C(X))$.

Proof. Fix any point $x \in X$ and let $S(x) := \{\varepsilon_x(S) : S \in S\}$, where

$\varepsilon_x : \Gamma(p) \to E_x$ is the evaluation map. As ε_x is continuous, the family $S(x)$ consists of compact, convex and circled sets and as $p : E \to X$ is a full bundle, this family covers E'_x. Therefore the $S(x)$-topology on E'_x is finer than the weak-$*$-topology and coarser than the Mackey--topology $\tau(E'_x, E_x)$. Thus, by the Mackey-Arens theorem, every continuous linear form on E'_x, equipped with the $S(x)$-topology, is of the form $\phi \to \phi(\alpha)$, where $\alpha \in E_x$.

As in section 11, we let $N_x^{Mod} = \{T \in Mod_S(\Gamma(p), C_b(X)) : T(\sigma)(x) = 0$ for all $\sigma \in \Gamma(p)\}$. Applying (11.6) we conclude that the set $\{f \cdot T : T \in Mod(\Gamma(p), C_b(X)),\ f \in C_b(X),\ f(x) = 0\}$ is dense in N_x^{Mod}. Moreover, if Φ is a $C_b(X)$-module homomorphism on $Mod_S(\Gamma(p), C_b(X))$, we have $\Phi(f \cdot T)(x) = (f \cdot \Phi(T))(x) = f(x) \cdot \Phi(T) = 0$ whenever $f(x) = 0$. Thus, the continuity of Φ implies $\Phi(T)(x) = 0$ for all $T \in N_x^{Mod}$. Let $\pi : Mod_S(\Gamma(p), C_b(X)) \to Mod_S(\Gamma(p), C_b(X))/N_x^{Mod}$ be the canonical quotient map. Then Φ induces a continuous map

$$\Phi_x : Mod(\Gamma(p), C_b(X))/N_x^{mod} \to \mathbb{K}$$

such that $\Phi_x \circ \pi(T) = \Phi(T)(x)$ for all $T \in Mod_S(\Gamma(p), C_b(X))$. By the remarks following (11.20), $Mod_S(\Gamma(p), C_b(X))/N_x^{Mod}$ may be identified with a subspace of $L_{S(x)}(E_x, \mathbb{K})$, i.e. with a subspace of E'_x equipped with the $S(x)$-topology. Under this identification, we have $\pi(T) = \lambda_T(x)$ and Φ_x becomes a continuous linear functional on a subspace of E'_x. Using the Hahn-Banach theorem, we may extend Φ_x to a continuous linear functional on E'_x, where E'_x carries the $S(x)$-topology and by the above remarks, there is an element $s(x) \in E_x$ such that $\Phi_x(\phi) = \phi(s(x))$ for all $\phi \in E'_x$.

Hence we have $\Phi(T)(x) = \Phi_x \circ \pi(T) = \Phi_x(\lambda_T(x)) = \lambda_T(x)(s(x))$ for all $T \in Mod_S(\Gamma(p), C_b(X))$ and $x \to s(x) : X \to E$ is the selection we were looking for. □

20.3 Remark. It is obvious from the proof of (20.2) that the section s in unique if and only if $\{\lambda_T(x) : T \in \mathrm{Mod}(\Gamma(p), C_b(X))\}$ is dense in E'_x for every $x \in X$.

Thus, the theorem of Mackey-Arens holds "internally" in the category of $C_b(X)$-modules, if we can show that the selection $s : X \to E$ constructed in (20.2) turns out to be continuous and bounded. For separable bundles of Banach spaces with continuous norm this is true. To obtain a more general result, we need some remarks concerning equivalent norms:

20.4 Definition. Let $p : E \to X$ be a fibered vector space. Two norms $||\cdot|| : E \to \mathbb{R}$ and $|||\cdot||| : E \to \mathbb{R}$ are said to be *equivalent*, if there are constants $m, M > 0$ such that $m\cdot ||\alpha|| \leq |||\alpha||| \leq M\cdot ||\alpha||$ for all $\alpha \in E$. □

20.5 Proposition. *Let* $p : E \to X$ *be a bundle of Banach spaces over a compact base space and with norm* $||\cdot||$. *Moreover, let* $|||\cdot|||$ *be a second norm on* E. *Then the following statements are equivalent:*

(i) The norms $||\cdot||$ and $|||\cdot|||$ are equivalent and the mapping $|||\cdot||| : E \to \mathbb{R}$ is upper semicontinuous.

(ii) $p : E \to X$ is a bundle of Banach spaces with norm $|||\cdot|||$.

(iii) The set

$$G = \bigcup_{x \in X} \{\alpha \in E_x : |||\alpha||| \leq 1\}^{\circ} \subset \bigcup_{x \in X} E'_x \subset \Gamma(p)'$$

is compact with respect to the weak-$*$-topology and the closed convex hull $G^{\circ\circ}$ of G is a barrel in $\Gamma(p)'$.

Proof. (i) → (ii): Firstly, we show that the set $\{\alpha \in E : |||\alpha||| < \varepsilon\}$ is open in E:

Let m,M > O such that $m\cdot\|\alpha\| \le |||\alpha||| \le M\cdot\|\alpha\|$ and let $\alpha_o \in E$ such that $|||\alpha_o||| < \varepsilon$. Choose any $\sigma \in \Gamma(p)$ with $\sigma(p(\alpha_o)) = \alpha_o$ and let

$$U := \{x \in X : |||\sigma(x)||| < \frac{\varepsilon + |||\alpha_o|||}{2}\}.$$

Then U is an open neighborhood of $p(\alpha_o)$ and thus

$$\mathcal{O} := \{\alpha \in E : p(\alpha) \in U \text{ and } \|\alpha - \sigma(p(\alpha))\| < \frac{\varepsilon - |||\alpha_o|||}{2\cdot M}\}$$

is an open neighborhood of α_o. Moreover, if α belongs to $\mathcal{O}$, then we have

$$\begin{aligned} |||\alpha||| &\le |||\sigma(p(\alpha)) - \alpha||| + |||\sigma(p(\alpha))||| \\ &< \frac{\varepsilon - |||\alpha_o|||}{2} + \frac{\varepsilon + |||\alpha_o|||}{2} \\ &= \varepsilon. \end{aligned}$$

It is now easy to check that $p : E \to X$ satisfies the hypothesis of (3.2) and thus is a bundle.

(ii) → (iii) : The compactness of G follows from (15.3) and by (15.7(i)) the closed convex hull of G is equal to the dual ball $\{\sigma \in \Gamma(p) : |||\sigma||| = \sup\{|||\sigma(x)||| : x \in X\} \le 1\}^o \subset \Gamma(p)'$ and therefore a barrel.

(iii) → (i) : As G is compact, we can find a constant M > O such that $G \subset \{\phi \in \Gamma(p)' : \|\phi\| \le M\}$ and as G^{oo} is a barrel, there is a constant m > O such that $\{\phi \in \Gamma(p)' : \|\phi\| \le m\} \subset G^{oo}$. From (15.15) applied to the bundle $p : E \to X$ equipped with the norm we conclude that $G^{oo} \cap E_x' = \{\phi \in E_x' : |||\phi||| \le 1\}$. Hence for every $x \in X$ we have the inclusions

$$\begin{aligned} \{\phi \in E_x' : \|\phi\| \le m\} &\subset \{\phi \in E_x' : |||\phi||| \le 1\} \\ &\subset \{\phi \in E_x' : \|\phi\| \le M\}. \end{aligned}$$

If we take polars, we obtain

$$\{\alpha \in E : \|\alpha\| \le \frac{1}{m}\} \subset \{\alpha \in E : |||\alpha||| \le 1\}$$
$$\subset \{\alpha \in E : \|\alpha\| \le \frac{1}{M}\}$$

or $M\cdot\|\alpha\| \le |||\alpha||| \le m\cdot\|\alpha\|$ for all $\alpha \in E$. It remains to show that the mapping $|||\cdot||| : E \to \mathbb{R}$ is upper semicontinuous, or, equivalently, that the mapping $x \to |||\sigma(x)||| : X \to \mathbb{R}$ is upper semicontinuous for every $\sigma \in \Gamma(p)$.

Thus, let $\sigma \in \Gamma(p)$. It is enough to prove that the set $\{x \in X : |||\sigma(x)||| \ge M\}$ is closed in X, where $M > 0$ is defined as above. Firstly, note that $|||\sigma(x)||| \ge M$ if and only if $|\phi(\sigma(x))| \ge 1$ for an appropriate element $\phi \in E'_x$ with $|||\phi||| \le \frac{1}{M}$. If we let

$$A_\sigma = \{\sigma \in \Gamma(p)' : |\phi(\sigma)| \ge 1\},$$

then we have

$$\{x \in X : |||\sigma(x)||| \ge M\} = \{x \in X : \tfrac{1}{M}\cdot G \cap E'_x \cap A_\sigma \neq \emptyset\}.$$

By the definition of M, the set $\frac{1}{M}\cdot G$ is contained in B_X and therefore $\frac{1}{M}\cdot G \cap A_\sigma$ is a compact subset of B_X not containing 0. Since the mapping

$$\gamma : B_X \setminus \{0\} \to X$$
$$\phi \to x \text{ iff } \phi \in E'_x$$

is continuous by (15.4), the image of $\frac{1}{M}\cdot G \cap A_\sigma$ under γ is compact in X and therefore closed. Since this image is exactly the set $\{x \in X: |||\sigma(x)||| \ge M\}$, our proof is complete. □

20.6 Proposition. *Let* $p : E \to X$ *be a bundle of Banach spaces over a compact base space* X *and with norm* $\|\cdot\|$. *If* $T \subset \Gamma(p)'$ *is a barrel, then the mapping*

$$|||\cdot||| : E \to \mathbb{R}$$
$$\alpha \to \sup\{|\phi(\alpha)| : \phi \in T \cap E'_{p(\alpha)}\}$$

is an upper semicontinuous norm on E *which is equivalent to* $\|\cdot\|$. *Moreover, if for each weak-$*$-open subset* $U \subset \Gamma(p)'$ *the set* $\{x \in X : U \cap E'_x \cap T \neq \emptyset\}$ *is open in* X, *then* $|||\cdot||| : E \to \mathbb{R}$ *is continuous.*

Proof. As $T \cap E'_x$ is a barrel in E'_x, the mapping $|||\cdot|||$ induces a norm on E_x, i.e. the mapping $|||\cdot||| : E \to \mathbb{R}$ is indeed a norm. Moreover, for every $x \in X$ we have

$$\{\alpha \in E_x : |||\alpha||| \leq 1\}^{\circ} = T \cap E'_x,$$

i.e. the set G defined in (20.5(iii)) is equal to $\bigcup_{x \in X} T \cap E'_x$.
Let $r > 0$ be a constant such that $r \cdot T \subset \{\phi \in \Gamma(p)' : \|\phi\| \leq 1\}$. Then $r \cdot T$ is a weak-$*$-compact subset of $\Gamma(p)'$ and hence $r \cdot T \cap B_X$ is compact, too. As $G = \frac{1}{r}(r \cdot T \cap B_X)$, the set G is also compact. Further let $s > 0$ be a real number such that $\{\phi \in \Gamma(p)' : \|\phi\| \leq 1\} \subset s \cdot T$. Then we may conclude that $B_X \subset s \cdot G$ and therefore

$$\begin{aligned}\{\phi \in \Gamma(p)' : \|\phi\| \leq 1\} &= B_X^{\circ\circ} \\ &\subset (s \cdot G)^{\circ\circ} \\ &= s \cdot G^{\circ\circ} \\ &\subset s \cdot T.\end{aligned}$$

This shows that $G^{\circ\circ}$ is a barrel.
The last statement follows from (15.11). □

The following corollary may be viewed as a complement to (19.16):

20.6 Corollary. *Let* $p : E \to X$ *be a bundle of Banach spaces and assume that the base space* X *is compact. If the set*

$$K = \{\phi \in B_X : \lambda_T(x) = \phi \text{ for some } x \in X \text{ and some } T \in Mod(\Gamma(p), C(X)) \text{ with } \|T\| \leq 1\}$$

is closed and if $\overline{\text{conv}}\ K$ *is a barrel, then there is a continuous norm* $|||\cdot||| : E \to \mathbb{R}$ *which is equivalent to* $\|\cdot\|$.

Proof. From (15.15) we conclude that $K = \bigcup_{x \in X} (\overline{\text{conv}}\ K) \cap E_x'$. Thus (20.6) follows from (20.5), as (19.12) yields that for every open set $U \subset \Gamma(p)'$ the set $\{x \in X : K \cap E_x' \cap U \neq \emptyset\}$ is open. □

We now come to the following theorem of the Mackey-Arens type:

20.7 Theorem. *Let* $p : E \to X$ *be a bundle of Banach spaces over a compact base space* X *and assume that there is a closed subset* K *of* $\Gamma(p)'$ *such that*

$$0 \in K \subset \{\phi \in B_X : \phi = \lambda_T(x) \text{ for some } x \in X \text{ and some } T \in Mod(\Gamma(p), C(X)) \text{ with } \|T\| \leq 1\}$$

Assume moreover that $E_x' \cap K$ *is convex and circled for every* $x \in X$ *and that* $\overline{\text{conv}}\ K$ *is a barrel in* $\Gamma(p)'$.
Then the mapping

$$\tilde{} : \Gamma(p) \to Mod(Mod_S(\Gamma(p), C(X)), C(X))$$
$$\sigma \to \tilde{\sigma} ; \quad \tilde{\sigma}(T) := T(\sigma)$$

is an isomorphism of $C(X)$*-modules, where* S *denotes any family of compact subsets of* $\Gamma(p)$ *such that the union of* S *generates* $\Gamma(p)$.

Proof. From (15.15) we know that $K \cap E_x' = (\overline{\text{conv}}\ K) \cap E_x'$, i.e. $K \cap E_x'$ is a barrel in E_x'. The injectivity of $\tilde{}$ now follows easily from (20.3).

Let $\phi : \mathrm{Mod}_S(\Gamma(p), C(X)) \to C(X)$ be a continuous $C(X)$-module homomorphism. Since in the Banach space $\Gamma(p)$ the closed convex hull of a compact subset is compact, Φ is also continuous for the finer topology of compact convex convergence on $\mathrm{Mod}(\Gamma(p), C(X))$. From (20.2) we may now conclude that there is a selection $s : X \to E$ such that $\Phi(T)(x) = \lambda_T(x)(s(x))$ for all $x \in X$ and all $T \in \mathrm{Mod}(\Gamma(p), C(X))$. It remains to show that s is continuous.

Firstly, by (20.6) we may assume without loss of generality that

$$\begin{aligned} \|\alpha\| &= \sup \{|\phi(\alpha)| : \phi \in \overline{\mathrm{conv}}\, K \cap E'_{p(\alpha)}\} \\ &= \sup \{|\phi(\alpha)| : \phi \in K \cap E'_{p(\alpha)}\} \end{aligned}$$

and that under these conditions we have $K = \mathcal{B}_X$. Further, by the choice of K, for every $\phi \in \mathcal{B}_X$ we can find a continuous $C(X)$-module homomorphism $T \in \mathrm{Mod}(\Gamma(p), C(X))$ with $\lambda_T(x) = \phi$, although we are no longer allowed to assume that $\|T\| \leq 1$.

Define a mapping

$$\tilde{s} : \bigcup_{x \in X} E'_x \to \mathbb{K}$$
$$s \to \tilde{s}\ ; \quad \tilde{s}(\phi) = \phi(s(x)) \text{ if } \phi \in E'_x.$$

<u>Step 1</u> The restriction of $\tilde{s}$ to $\mathcal{B}_X$ is continuous.
(Let $\mathcal{M}^p_S := \{(x,\phi) : \phi = \lambda_T(x)$ for some $T \in \mathrm{Mod}(\Gamma(p), C(X))\} \subset X \times \Gamma_S(p)'$ and let $\pi^p : \mathcal{M}^p_S \to X$ be the restriction of the first projection. As we just remarked, $\{x\} \times E'_x \subset \mathcal{M}^p_S$. By (19.2), $\pi^p : \mathcal{M}^p_S \to X$ is a bundle and the mapping

$$\Lambda : \mathrm{Mod}\ (\Gamma(p), C(X)) \to \Gamma(\pi^p)$$
$$T \to \Lambda(T)\ ; \quad \Lambda(T)(x) = (x, \lambda_T(x))$$

is a topological isomorphism of $C(X)$-modules. As $\Phi \circ \Lambda^{-1} : \Gamma(\pi^p) \to C(X)$

is a continuous $C(X)$-module homomorphism, we can find an unique bundle morphism $\mu : M_S^p \to X \times \mathbb{K}$ such that

$$(\Phi \circ \Lambda^{-1})(\chi)(x) \quad = \quad (pr_2 \circ \mu \circ \chi)(x) \qquad \text{for all } x \in X,\ \chi \in \Gamma(\pi^p).$$

As μ preserves stalks and as the stalks of π^p are just the $\{x\} \times E_x'$, $x \in X$, we can find a linear map $\mu_x : E_x' \to \mathbb{K}$ such that $\mu(x,\phi) =$ $= (x, \mu_x(\phi))$ for every $\phi \in E_x'$.

Now suppose that $T \in \mathrm{Mod}(\Gamma(p), C(X))$. Then we may compute:

$$\begin{aligned}
\lambda_T(x)(s(x)) &= \Phi(T)(x) \\
&= \Phi \circ \Lambda^{-1}(\Lambda(T))(x) \\
&= pr_2 \circ \mu(\Lambda(T)(x)) \\
&= pr_2 \circ \mu(x, \lambda_T(x)) \\
&= pr_2(x, \mu_x(\lambda_T(x))) \\
&= \mu_x(\lambda_T(x)).
\end{aligned}$$

As the elements of the form $\lambda_T(x)$, $T \in \mathrm{Mod}(\Gamma(p), C(X))$ cover E_x', we have

$$\begin{aligned}
pr_2 \circ \mu(x,\phi) &= \mu_x(\phi) \\
&= \phi(s(x)) \\
&= \tilde{s} \circ pr_2(x,\phi) \qquad\qquad \text{for all } \phi \in E_x'.
\end{aligned}$$

Restricting μ to the set $\{(x,\phi) : \phi \in B_x\} \subset X \times B_X$ we obtain the following commutative diagram

$$\begin{array}{ccc}
\{(x,\phi) : \phi \in B_x,\ x \in X\} & \xrightarrow{\ \mu\ } & X \times \mathbb{K} \\
pr_2 \downarrow & & \downarrow pr_2 \\
B_X & \xrightarrow[\ \tilde{s}\ ]{} & \mathbb{K}
\end{array}$$

It follows easily from (15.4) that the set $\{(x,\phi) : \phi \in B_x,\ x \in X\}$

is compact, if we equip $\mathcal{B}_X$ with the weak-$*$-topology. Since the subset $\mathcal{B}_X \subset \Gamma(p)'$ is equicontinuous, the weak-$*$-topology and the $\mathcal{S}$-topology on $\mathcal{B}_X$ agree. Thus the set $\{(x,\phi) : \phi \in \mathcal{B}_x, x \in X\}$ is compact in the relative topology of M_S^p and therefore the projection $pr_2 : \{(x,\phi) : \phi \in \mathcal{B}_x, x \in X\} \to \mathcal{B}_X$ is a quotient map as it is a continuous surjection between compact spaces. As μ and $pr_2 : X \times \mathbb{K} \to \mathbb{K}$ are continuous, the continuity of $\tilde{s}$ follows.)

<u>Step 2</u>. If $s(x_o) = 0$, then the set $\{x : \|s(x)\| < \varepsilon\}$ is a neighborhood of x_o.

(Let

$$U := \{\phi \in \mathcal{B}_X : |\tilde{s}(\phi)| < \varepsilon/2\}.$$

Then U is open by step 1. Moreover, $s(x_o) = 0$ implies $\tilde{s}(\phi) = \phi(s(x_o)) = 0$ for all $\phi \in \mathcal{B}_{x_o}$, i.e. $\mathcal{B}_{x_o} \subset U$. Hence the set $\{x \in X : \mathcal{B}_x \subset U\}$ is an open neighborhood of x_o by (15.6). As for every $x \in X$ with $\mathcal{B}_x \subset U$ we have $\|s(x)\| = \sup\{|\phi(s(x))| : \phi \in \mathcal{B}_x\} = \sup\{|\tilde{s}(\phi)| : \phi \in \mathcal{B}_x\} \le \varepsilon/2 < \varepsilon$, the larger set $\{x \in X : \|s(x)\| < \varepsilon\}$ is a neighborhood of x_o, too.)

<u>Step 3</u>. The mapping $s : X \to E$ is continuous.

(Let $x_o \in X$ and let $\mathcal{O}$ be an open neighborhood of $s(x_o)$. Pick any section $\sigma \in \Gamma(p)$ such that $\sigma(x_o) = s(x_o)$. Then there are an open neighborhood U of x_o and a real number $\varepsilon > 0$ such that

$$\{\alpha \in E : p(\alpha) \in U \text{ and } \|\alpha - \sigma(p(\alpha))\| < \varepsilon\} \subset \mathcal{O}.$$

To show the continuity of s, it is enough to check that the set

$$\{x \in X : \|\sigma(x) - s(x)\| < \varepsilon\}$$

is a neighborhood of x_o.

Let $\Psi = \tilde{\sigma} - \Phi : \mathrm{Mod}(\Gamma(p), C(X)) \to C(X)$. Then there is an unique selection $r : X \to E$ so that $\Psi(T)(x) = \lambda_T(x)(r(x))$ for all $x \in X$ and all $T \in \mathrm{Mod}(\Gamma(p), C(X))$. An easy computation shows that $r = \sigma - s$ and therefore step 2 applied to r instead of s completes the proof. □

20.8 Corollary. *Let* $p : E \to X$ *be a separable bundle of Banach spaces with continuous norm and a compact base space. Then the mapping*

$$\tilde{} \quad : \quad \Gamma(p) \to \mathrm{Mod}(\mathrm{Mod}_S(\Gamma(p), C(X)), C(X))$$
$$\sigma \to \tilde{\sigma}$$

is a bijection. □

Our next corollary deals with locally trivial bundles. Firstly, however, we need a lemma:

20.9 Lemma. *Let* $p : E \to X$ *be a bundle of Banach spaces over a compact base space* X *and suppose that the bundle* $p : E \to X$, *viewed as a bundle of topological spaces, is locally trivial. Then there is a compact subset* $A \subset \mathrm{Mod}_s(\Gamma(p), C(X))$ *such that*

(i) $K = \{\lambda_T(x) : x \in X \text{ and } T \in A\}$ *is compact*

(ii) $0 \in K$ *and* $K \cap E'_x$ *is circled and convex for every* $x \in X$.

(iii) $\overline{\mathrm{conv}}\, K$ *is a barrel in* $\Gamma(p)'$.

Proof. Let $A_1, \ldots, A_n$ be closed subsets of X such that the interiors of the A_i cover X and such that $\Gamma_{A_i}(p)$ is isomorphic to $C(A, E)$ as a topological vector space for a certain Banach space E. Since the restriction map

$$\varepsilon_{A_i} \quad : \quad \Gamma(p) \to \Gamma_{A_i}(p)$$
$$\sigma \to \sigma/_{A_i}$$

is a quotient map by (4.5), we may embed $C(A_i,E)'$ into $\Gamma(p)'$ via the mapping

$$e_i : C(A_i,E)' \to \Gamma(p)'$$
$$e_i(\phi)(\sigma) = \phi \circ S_i \circ \varepsilon_{A_i}(\sigma)$$

where $S_i : \Gamma_{A_i}(p) \to C(A_i,E)$ is a suitable continuous and open $C(X)$--module isomorphism. For every $i \in \{1,\dots,n\}$ and every $\phi \in E'$ we define a mapping

$$\eta_\phi : A_i \to C(A_i,E)' ,$$

where

$$\eta_\phi(x)(\tau) = \phi(\tau(x)) \qquad \text{for all } \tau \in C(A_i,E).$$

It is straightforward to check that η_ϕ is continuous and that $e_i \circ \eta_\phi$ maps x into E'_x, where, as usual, $E_x = p^{-1}(x)$.
For every $i \in \{1,\dots,n\}$ let $f_i : X \to [0,1]$ be a continuous function such that f_i vanishes on $X \setminus A_i^o$ and such that

$$\max_{1\le i\le n} f_i(x) = 1 \qquad \text{for all } x \in X.$$

If we define

$$\lambda_{\phi,i} : X \to \Gamma_s(p)'$$

by

$$\lambda_{\phi,i}(x) = \begin{cases} f_i(x)\cdot(e_i \circ \eta_\phi)(x) & \text{if } x \in A_i \\ 0 & \text{if } x \notin A_i \end{cases}$$

then $\lambda_{\phi,i}$ is continuous and $\lambda_{\phi,i}(x) \in E'_x$ for every $x \in X$.
For every $1 \le i \le n$ we define a mapping

$$m_i : E' \to Mod_s(\Gamma(p),C(X))$$
$$\phi \to T_{\lambda_{\phi,i}}.$$

An easy calculation shows that m_i is linear. Moreover, we have

$$\begin{aligned}\|T_{\lambda_{\phi,i}}(\sigma)\| \le 1 \quad &\text{iff} \quad |\lambda_{\phi,i}(x)(\sigma(x))| \le 1 && \text{for all } x \in X\\ &\text{iff} \quad |f_i(x)(e_i \circ \eta_\phi(x))(\sigma(x))| \le 1 && \text{for all } x \in A_i\\ &\text{iff} \quad |(e_i \circ \eta_\phi(x))((f_i \cdot \sigma)(x))| \le 1 && \text{for all } x \in A_i\\ &\text{iff} \quad |\phi((S_i \circ \varepsilon_{A_i}(f_i \cdot \sigma))(x))| \le 1 && \text{for all } x \in A_i.\end{aligned}$$

As the set $\{[(S_i \circ \varepsilon_{A_i})(f_i \cdot \sigma)](x) : x \in A_i\}$ is compact in E, we conclude that m_i is continuous if E' carries the topology of compact convergence. As the set $\{\phi \in E' : \|\phi\| \le 1\}$ is compact in this topology, the image

$$\mathcal{B}_i := \{T_{\lambda_{\phi,i}} : \|\phi\| \le 1\}$$

of the unit ball of E' under m_i is compact, too.

Now let $\mathcal{B} = \mathcal{B}_1 \cup \ldots \cup \mathcal{B}_n$ and let A be the closed, convex, circled hull of $\mathcal{B}$. As $\Gamma(p)$ and $C(X)$ are Banach spaces, the space $L_s(\Gamma(p),C(X))$ of all linear operators from $\Gamma(p)$ into $C(X)$ is quasicomplete and so is its closed subspace $Mod_s(\Gamma(p),C(X))$. Thus, A is a compact convex and circled subset of $Mod_s(\Gamma(p),C(X))$.
Obviously, $0 = T_{\lambda_{0,i}} \in A$.

Let

$$K = \{\lambda_T(x) : T \in A,\ x \in X\}.$$

Then, by definition, for every $i \in \{1,\ldots,n\}$ and every $\phi \in E'$ we have $\lambda_{\phi,i} \in K$, $0 \in K$ and $K \cap E_x$ is convex and circled for every $x \in X$. Moreover, the set K is compact: Since A is compact, it is enough to show that the mapping $(x,T) \to \lambda_T(x) : X \times A \to K$ is continuous.

Let $(x_i, T_i)_{i \in I}$ be a converging net in $X \times A$ and let $x = \lim\limits_{i \in I} x_i$ and let $T = \lim\limits_{i \in I} T_i$. We show:

$$\lambda_T(x) = \lim_{i \in I} \lambda_{T_i}(x_i).$$

Indeed, let $\sigma \in \Gamma(p)$. As $\lim\limits_{i \in I} T_i = T$, there is a $j_1 \in I$ such that $\|T_i(\sigma) - T(\sigma)\| < \varepsilon/2$ for all $i \geq j_1$. Moreover, as $T(\sigma) \in C(X)$, there is a $j_2 \geq j_1$ such that $|T(\sigma)(x) - T(\sigma)(x_i)| < \varepsilon/2$ for all $i \geq j_2$. For all $i \geq j_2 \geq j_1$ we have

$$\begin{aligned} |\lambda_{T_i}(x_i)(\sigma) - \lambda_T(x)(\sigma)| &= |T_i(\sigma)(x_i) - T(\sigma)(x)| \\ &\leq |T_i(\sigma)(x_i) - T(\sigma)(x_i)| + |T(\sigma)(x_i) - \\ &\qquad - T(\sigma)(x)| \\ &\leq \|T_i(\sigma) - T(\sigma)\| + \varepsilon/2 \\ &\leq \varepsilon/2 + \varepsilon/2 = \varepsilon. \end{aligned}$$

It remains to show that $\overline{\text{conv}}\, K = K^{oo}$ is a barrel, i.e. that K^o is bounded in $\Gamma(p)$.

For $1 \leq i \leq n$ let

$$B_i = \{x \in X : f_i(x) = 1\}.$$

Then $B_i \subset A_i^o$ and the B_i cover X as we have chosen the f_i so that $\max\{f_i(x) : 1 \leq i \leq n\} = 1$ for all $x \in X$.

Let $\delta_i : \Gamma_{A_i}(p) \to \Gamma_{B_i}(p)$ and $\delta_i' : C(A_i, E) \to C(B_i, E)$ be the restriction maps. Then δ_i and δ_i' are quotient maps by (4.5). Moreover, there is a (topological) isomorphism $R_i : \Gamma_{B_i}(p) \to C(B_i, E)$ such that the diagram

$$\begin{array}{ccccc} \Gamma(p) & \xrightarrow{\varepsilon_{A_i}} & \Gamma_{A_i}(p) & \xrightarrow{S_i} & C(A_i, E) \\ \varepsilon_{B_i} \downarrow & & \downarrow \delta_i & & \downarrow \delta_i' \\ \Gamma_{B_i}(p) & \xrightarrow{id} & \Gamma_{B_i}(p) & \xrightarrow{R_i} & C(B_i, E) \end{array}$$

commutes for every $i \in \{1,\dots,n\}$.

Given $x \in B_i$ and $\sigma \in K^o$, we compute

$$
\begin{aligned}
\|((R_i \circ \varepsilon_{B_i})(\sigma))(x)\| &= \|((\delta_i' \circ S_i \circ \varepsilon_{A_i})(\sigma))(x)\| \\
&= \|((S_i \circ \varepsilon_{A_i})(\sigma))(x)\| \quad \text{by the definition of } \delta_i' \\
&= \sup\{|\phi(((S_i \circ \varepsilon_{A_i})(\sigma))(x))| : \phi \in E',\ \|\phi\| \le 1\} \\
&= \sup\{|\eta_\phi(x)((S_i \circ \varepsilon_{A_i})(\phi))| : \phi \in E',\ \|\phi\| \le 1\} \\
&\qquad \text{by the definition of } \eta_\phi \\
&= \sup\{|(e_i(\eta_\phi(x)))(\sigma)| : \phi \in E',\ \|\phi\| \le 1\} \\
&\qquad \text{by the definition of } e_i \\
&= \sup\{|\lambda_{\phi,i}(x)(\sigma) : \phi \in E',\ \|\phi\| \le 1\} \\
&\qquad \text{since } f_i(x) = 1 \text{ on } B_i \\
&\le 1 \qquad \text{since } \lambda_{\phi,i} \in K \text{ and as } \sigma \in K^o
\end{aligned}
$$

Thus, we conclude that $\|R_i \circ \varepsilon_{B_i}(\sigma)\| \le 1$ for every $\sigma \in K^o$. Since R_i is a topological isomorphism, there is a constant $M_i > 0$ such that $\|\varepsilon_{B_i}(\sigma)\| \le M_i$ for every $\sigma \in K^o$. Let

$$M = \max\{M_1,\dots,M_n\}.$$

As the B_i cover X, we conclude that $\|\sigma\| \le M$ for every $\sigma \in K^o$ and our proof is complete. □

20.10 Corollary. *Let* $p : E \to X$ *be a bundle of Banach spaces over a compact base space and assume that* $p : E \to X$, *viewed as a bundle of topological vector spaces, is locally trivial. Then the mapping*

$$\tilde{} : \Gamma(p) \to \mathrm{Mod}(\mathrm{Mod}_S(\Gamma(p), C(X)), C(X))$$

is a bijection, where S denotes any family of compact subsets of $\Gamma(p)$ *such that the union of S generates* $\Gamma(p)$. □

For bundles with finite dimensional stalks we have the following result:

20.11 Corollary. *Let* $p : E \to X$ *be a Banach bundle with a compact base space* X *and assume that all stalks are finite dimensional.*

(i) *If the mapping* $\tilde{}$: $\Gamma(p) \to \mathrm{Mod}(\mathrm{Mod}_S(\Gamma(p),C(X)),C(X))$ *is bijective, then* E *is a Hausdorff space. Further, for a given* $x \in X$ *and a given* $\phi \in E_x'$ *there is a continuous* C(X)*-module homomorphism* $T \in \mathrm{Mod}(\Gamma(p),C(X))$ *such that* $\lambda_T(x) = \phi$.

(ii) *Conversely, assume that* E *is Hausdorff. If the base space* X *is metrizable* <u>*or*</u> *if there is an* $n \in \mathbb{N}$ *such that* $\dim E_x = n$ *for all* $x \in X$*, then the mapping* $\tilde{}$ *is a bijection.*

Proof. (i) From (20.3) and the fact that E_x' is finite dimensional for every $x \in X$ it follows that $E_x' = \{\lambda_T(x) : T \in \mathrm{Mod}(\Gamma(p),C(X))\}$. Now (19.3(ii)) yields that E is Hausdorff.

(ii) If X is metrizable, then $p : E \to X$ is separable by (19.5(iii)) and therefore $\tilde{}$ is a bijection by (20.8).

Now suppose that $\dim E_x = n$ for all $x \in X$, where $n \in \mathbb{N}$ is fixed. Then $p : E \to X$ is locally trivial by (18.5). In this case, (ii) follows from (20.10). □

20.12 Definition. Let $p : E \to X$ be a bundle. Then p is called a *Mackey bundle*, provided that the mapping $\tilde{}$: $\Gamma(p) \to \mathrm{Mod}_{cc}(\mathrm{Mod}_{cc}(\Gamma(p), C(X)),C(X))$ is a homeomorphism, where the subscript "cc" refers to the topology of uniform convergence on compact, convex circled subsets. □

20.13 Remarks. It is easy to see that $\tilde{}$ is continuous whenever every compact convex and circled subset of $\mathrm{Mod}(\Gamma(p),C(X))$ is equicontinuous. This is especially the case if $p : E \to X$ is a bundle of Banach spaces.

Before we give a very meager set of examples of Mackey bundles, we shall establish:

20.14 Proposition. *Let* $p : E \to X$ *be a bundle of Banach spaces over a compact base space* X. *Moreover, let* $A \subset Mod_s(\Gamma(p), C(X))$ *be a compact subset.*

(i) *The set* $K_A = \{\lambda_T(x) : x \in X, T \in A\}$ *is compact.*

(ii) $p : E \to X$ *is a Mackey bundle if and only if there is a compact subset* $A \subset Mod_s(\Gamma(p), C(X))$ *such that the closed convex circled hull of* K_A *is a barrel.*

Proof. (i) was already shown in the proof of (20.9).

(ii): Suppose that $p : E \to X$ is a Mackey bundle. Then the mapping $\tilde{} : \Gamma(p) \to Mod_{cc}(Mod_{cc}(\Gamma(p), C(X)), C(X))$ is open. Hence we can find a compact, convex and circled subset $A \subset Mod_{cc}(\Gamma(p), C(X))$ such that $||T(\sigma)|| \le 1$ for all $T \in A$ implies $||\sigma|| \le 1$. Clearly, the set A is also compact in $Mod_s(\Gamma(p), C(X))$ and it remains to show that K_A^o is bounded.

Thus, let $\sigma \in K_A^o$. We show that $||\sigma|| \le 1$. Indeed, as $\sigma \in K_A^o$, we know that $|T(\sigma)(x)| = |\lambda_T(x)(\sigma(x))| \le 1$ for all $x \in X$ and all $T \in A$, i.e. $||T(\sigma)|| \le 1$ for all $T \in A$ and therefore $||\sigma|| \le 1$ by the choice of A.

Conversely, suppose that $A \subset Mod_s(\Gamma(p), C(X))$ is given such that the closed convex circled hull of K_A is a barrel in $\Gamma(p)'$. We may suppose that A is circled. Hence the set K_A is circled, too, and the closed convex circled hull of K_A is equal to K_A^{oo}. Since K_A^{oo} is a barrel, there is a constant $M > 0$ such that $||\phi|| \le M$ implies $\phi \in K_A^{oo}$ for all $\phi \in \Gamma(p)'$. Now let us assume that $||T(\sigma)|| \le 1$ for all $T \in A$. Then we may conclude that $|\lambda_T(x)(\sigma(x))| \le 1$ for all $x \in X$ and all $T \in A$, i.e. $\sigma \in K_A^o$. This implies $|\phi(\sigma)| \le 1$ whenever

$||\phi|| \leq M$, i.e. $||\sigma|| \leq \frac{1}{M}$. Thus, we have shown that $||T(\sigma)|| \leq M$ for all $T \in A$ implies $||\sigma|| \leq 1$.
As every compact subset $A \subset Mod_s(\Gamma(p),C(X))$ is compact in the stronger topology $Mod_{cc}(\Gamma(p),C(X))$ and as the closed convex circled hull of A is also compact, we just verified the openess of $\tilde{}$. Since the map $\tilde{}$ is continuous by (20.13) and since it follows from (20.7) that $\tilde{}$ is bijective, the proof of (20.14) is complete. □

20.15 Examples. (i) If X is compact and if $p : E \to X$ is a locally trivial bundle of Banach spaces, then p is a Mackey bundle (see (20.9))and (20.10)).
(ii) If $p : E \to X$ is a bundle of Banach spaces, if X is compact, if E is Hausdorff and if all stalks have dimension n for a fixed $n \in \mathbb{N}$, then $p : E \to X$ is a Mackey bundle (see (18.5), (20.9), (20.10)).

20.16 Proposition. *Let* $p : E \to X$ *be a Mackey bundle of Banach spaces, where* X *is compact. Then for every* $x \in X$ *and every* $\phi \in E'_x$ *there is a continuous* C(X)*-module homomorphism* $T \in Mod(\Gamma(p),C(X))$ *such that* $\lambda_T(x) = \phi$*, i.e. the stalks of the bundle* $\pi^p : M^p_S \to X$ *representing* $Mod_S(\Gamma(p),C(X))$ *are isomorphic to* E'_x, $x \in X$, *where* S *denotes again a family of compact subsets of* $\Gamma(p)$ *whose union generates* $\Gamma(p)$.

Proof. By (20.14) there is a compact subset $A \subset Mod_s(\Gamma(p),C(X))$ such that the closed convex circled hull of $K_A = \{\lambda_T(x) : x \in X$ and $T \in A\}$ is a barrel in $\Gamma(p)'$. As $\Gamma(p)$ and $C(X)$ are Banach spaces, the closed convex circled hull of A is compact, too. Hence we may assume w.l.o.g. that A is convex and circled. Moreover, by [Sch 71, III.4.2] the set A is equicontinuous. Therefore, by multiplying A with a suitable constant $M > 0$, we may assume that $||T|| \leq 1$ for all $T \in A$.
Under these conditions we have

(1) $K_A \subset B_X$ and K_A is closed

(2) $K_A \cap B_x$ is closed, convex and circled for every $x \in X$.

(3) $K_A^{oo} = \overline{\text{conv}}\, K_A$ is a barrel in $\Gamma(p)'$.

where the closedness of $K_A \cap B_x$ follows from (20.14(i)). Applying (15.15), we are allowed to conclude that $K_A \cap B_x = (\overline{\text{conv}}\, K_A) \cap B_x$ is a barrel in E_x' for every $x \in X$.

Now let $x \in X$ and let $\phi \in E_x'$ be given. Then there is a real number $r > 0$ such that $r\cdot\phi \in K_A \cap B_x$. By the definition of K_A we can find a continuous $C(X)$-module homomorphism $T' \in A$ such that $r\cdot\phi = \lambda_{T'}(x)$. Let $T := \frac{1}{r}\cdot T'$. Then we finally have $\phi = \lambda_T(x)$. □

20.17 Theorem. *If $p : E \to X$ is a Mackey bundle of Banach spaces over a compact base space with norm $\|\cdot\|$, then there is an equivalent norm $|||\cdot|||$ on E such that for every $x_o \in X$ and every $\phi \in E_{x_o}'$ there is a $T \in \text{Mod}(\Gamma(p),C(X))$ with $\lambda_T(x_o) = \phi$ and $|||\phi||| = |||T|||$. Furthermore, the mapping $|||\cdot||| : E \to \mathbb{R}$ is continuous.*

Proof. As in the proof of (20.16), let $A \subset \text{Mod}_s(\Gamma(p),C(X))$ be a compact, convex and circled subset such that

(1) $K_A \subset B_X$ and K_A is closed.

(2) $K_A \cap B_x$ is closed, convex and cirlced for every $x \in X$.

(3) $K_A^{oo} = \overline{\text{conv}}\, K_A$ is a barrel in $\Gamma(p)'$.

Again, proposition (15.15) implies

(4) $K_A \cap B_x = (\overline{\text{conv}}\, K_A) \cap B_x$ for each $x \in X$.

We now define a norm on E by

$$|||\alpha||| = \sup \{|\phi(\alpha)| : \phi \in K_A \cap B_{p(\alpha)}\}.$$

As it was shown in (20.6), the mapping $|||\cdot||| : E \to \mathbb{R}$ is an equivalent and upper semicontinuous norm on E. Moreover, as $K_A \cap B_x$ is convex, circled and closed, we have $|||\phi||| \le 1$ if and only if ϕ belongs to $\{\alpha \in E_x : |||\alpha||| \le 1\}^o$ if and only if $\phi \in (K_A \cap B_x)^{oo} = K_A \cap B_x$ for a certain $x \in X$.

Next, we show that $|||T||| \le 1$ for every $T \in A$. Indeed, if $\sigma \in \Gamma(p)$ is given such that $|||\sigma||| \le 1$, then we compute

$$\begin{aligned} |||T(\sigma)||| &= \sup\{|T(\sigma)(x)| : x \in X\} \\ &= \sup\{|\lambda_T(x)(\sigma(x))| : x \in X\} \\ &\le 1 \end{aligned}$$

as $\lambda_T(x) \in K_A \cap B_x$ and as $\sigma(x) \in (K_A \cap B_x)^o$.

Further, let us start with $x_o \in X$ and let $\phi \in E'_{x_o} \setminus \{0\}$. Then the element $|||\phi|||^{-1}\cdot\phi$ belongs to $K_A \cap B_{x_o}$ and therefore is of the form $\lambda_S(x_o)$ for a certain $S \in A$. Since $|||S||| \le 1$ by the above argument and since

$$\begin{aligned} |||S||| &= \sup\{|||\lambda_S(x)||| : x \in X\} \\ &\ge |||\lambda_S(x_o)||| \\ &= |||\phi||| \\ &= 1, \end{aligned}$$

by (10.13) (recall that λ_S may be considered as a bundle morphism between E and the trivial bundle $X\times\mathbb{K}$!), we conclude that $|||S||| = 1$. Now define $T \in \mathrm{Mod}(\Gamma(p), C(X))$ by $T := |||\phi|||\cdot S$. Then $\lambda_T(x_o) = \phi$ and $|||T||| = |||\phi|||$.

Finally, the continuity of $|||\cdot||| : E \to \mathbb{R}$ follows exactly as in the proof of (19.16, (ii) → (iii) → (i)). □

21. The dual space $\Gamma(p)'$ of a space of sections.

In the preceeding sections we always used the dual space $\Gamma(p)'$ of a space of sections in a bundle to construct subbundles, $C(X)$-module homomorphisms etc. In this last section, we would like to reverse these questions: Suppose that we already know the "intern dual" $Mod(\Gamma(p),C(X))$, what can be said about $\Gamma(p)'$ itself? Of course we can expect reasonable answers only if $Mod(\Gamma(p),C(X))$ is large enough to separate the points of $\Gamma(p)$ and it will turn out that we need more than this.

Our first observation is the following: Given a $C(X)$-module E, then the dual space E' is also a $C(X)$-module, if we define a multiplication on E' by

$$(f\cdot\phi)(\sigma) = \phi(f\cdot\sigma) \qquad \text{for all } f \in C_b(X), \sigma \in E, \phi \in E'$$

but there is no reason to expect that E' is locally $C(X)$-convex even when E has this property. An example for this phenomenon is $E = C(X)$ itself.

On the other hand, given a $C(X)$-module homomorphism $T : E \to C_b(X)$, we may map the dual space $M(X)$ of $C_b(X)$ into E' via the function

$$\begin{aligned} T\boxtimes - \; : \; M(X) &\to E' \\ \mu &\to T\boxtimes\mu \;, \quad T\boxtimes\mu(\sigma) = \mu(T(\sigma)) \end{aligned}$$

and in certain cases the images of this mapping will generate E'. In these cases, we shall obtain something close to a "integral representation" of linear functionals on E. If $p : E \to X$ is a bundle, a typical linear functional looks like

$$T \boxtimes \mu \ : \ \Gamma(p) \to \mathbb{K}$$
$$\sigma \to \int_X \lambda_T(x)(\sigma(x))\, d\mu$$

where $T \in Mod(\Gamma(p), C(X))$ and where $\mu \in M(X)$.

21.1 Definition. (i) Let E,F,G be topological vector spaces and let $b : E \times F \to G$ be a bilinear mapping. If b is separately continuous on $E \times F$ and if for every bounded subset $B \subset F$ the family of linear maps $b(-,u) : E \to G$, $u \in B$, is equicontinuous, then b is called *hypocontinuous*.
(ii) If E and F are $C_b(X)$-modules and if $b : E \times F \to G$ is bilinear and satisfies

$$b(u, f \cdot v) = b(f \cdot u, v) \quad \text{for all } u \in E,\ v \in F,\ f \in C_b(X)$$

then we say that b is *compatible with the* $C_b(X)$*-module structure*.
(iii) If in addition G is an C(X)-module, too, and if

$$b(u, f \cdot v) = b(f \cdot u, v) = f \cdot b(u,v) \quad \text{for all } u \in E,\ v \in F \text{ and } f \in C_b(X),$$

then we call b a *bilinear mapping between* $C_b(X)$*-modules*. □

In the following we denote the dual space of $C_b(X)$, where X is a topological space, by M(X).
Again, if E is a topological vector space and if S is any directed family of bounded subsets of E, we denote the topology of uniform convergence on elements of S defined on a space of mappings with domain E by adding the subscript S.

21.2 Proposition. *Let* E *be a topological* $C_b(X)$*-module, let* S *be a directed and total family of bounded subsets of* E *and let* S' *be* *be a directed family of bounded subset of* $C_b(X)$ *whose union generates* $C_b(X)$. *Assume that* $\{T(S) : S \in \mathcal{S}, T \in \mathrm{Mod}(E,C(X))\}$ *is contained in* $\mathcal{S}'$.

(i) *The mapping*

$$\boxtimes : \mathrm{Mod}_{\mathcal{S}}(E,C_b(X)) \times M_{\mathcal{S}'}(X) \to E'_{\mathcal{S}}$$
$$(T,\mu) \to T\boxtimes\mu;\ (T\boxtimes\mu)(\sigma) = \mu(T(\sigma))$$

is a hypocontinuous bilinear mapping between $C_b(X)$*-modules*

(ii) *If* $\mathcal{S}$ *covers* E, *and if* $\sigma \in E$ *is given, then the mapping*

$$b_\sigma : \mathrm{Mod}_{\mathcal{S}}(E,C_b(X))\ M_{\mathcal{S}'}(X) \to \mathbb{K}$$
$$(T,\mu) \to \mu(T(\sigma))$$

is hypocontinuous, bilinear and compatible with the $C_b(X)$*-module structure.*

(iii) *If* $\mathrm{Mod}(E,C_b(X))$ *separates the points of* E, *then the linear span of the image of* $\boxtimes$ *is* $\sigma(E',E)$*-dense in* E'.

Proof. (i) For all $f \in C_b(X)$, all $\sigma \in E$, all $\mu \in M(X)$ and all $T \in \mathrm{Mod}(E,C_b(X))$ we have

$$\begin{aligned}
((f\cdot T)\boxtimes\mu)(\sigma) &= \mu((f\cdot T)(\sigma)) \\
&= \mu(f\cdot(T(\sigma))) && \text{by the definition of the multiplication on } \mathrm{Mod}(E,C_b(X)) \\
&= (f\cdot\mu)(T(\sigma)) && \text{by the definition of the multiplication on } M(X) \\
&= (T\boxtimes(f\cdot\mu))(\sigma) \\
&= \mu(T(f\cdot\sigma)) && \text{as } T \in \mathrm{Mod}(E,C_b(X)) \\
&= (T\boxtimes\mu)(f\cdot\sigma) \\
&= (f\cdot(T\boxtimes\mu))(\sigma) && \text{by the definition of the multiplication on } E'
\end{aligned}$$

establishing the fact that $\boxtimes$ is a bilinear mapping between C(X)-modules.

If we fix $T \in \mathrm{Mod}(E,C_b(X))$, then

$$T\boxtimes - \; : M_{S'}(X) \to E'_S$$

is continuous. Indeed, if $S \in \mathcal{S}$, then the set $S' := T(S)$ belongs to $\mathcal{S}'$. Moreover, if $\mu \in M(X)$ belongs to the open neighborhood

$$\{\nu \in M(X) : |\nu(f)| \le 1 \text{ for all } f \in S'\}$$

of 0 in $M_{\mathcal{S}'}(X)$, then we have the inequality $|(T\,\mu)(\sigma)| = |\mu(T(\sigma))| \le$ ≤ 1 for all $\sigma \in S$. Since the set

$$\{\phi \in E' : |\phi(\sigma)| \le 1 \text{ for all } \sigma \in S\}$$

is a basic neighborhood of 0 in $E'_{\mathcal{S}}$, we have shown that the mapping $T\boxtimes -$ is continuous.

Now let $A \subset M_{\mathcal{S}'}(X)$ be bounded. As $\mathcal{S}'$ covers $C_b(X)$, the corollary to [Sch 71, III.3.4] yields an $M > 0$ such that $\|\mu\| \le M$ for all $\mu \in A$. Hence, if $S \in \mathcal{S}$ is given and if $T \in \mathrm{Mod}(E,C_b(X))$ satisfies $\|T(\sigma)\| \le \frac{1}{M}$ for all $\sigma \in E$, then for every $\sigma \in S$ and every $\mu \in A$ we have $|T\,\mu(\sigma)| = |\mu(T(\sigma))| \le \|\mu\|\cdot\|T(\sigma)\| \le M\cdot\frac{1}{M} = 1$ showing the equicontinuity of the set $\{-\boxtimes\mu : \mu \in A\}$.

The proofs of (ii) and (iii) are now straightforward. □

The following results state a converse of (21.2(ii)):

21.3 Proposition. *Let E be a topological $C_b(X)$-module and let $\mathcal{S}$ and $\mathcal{S}'$ resp. be directed and covering families of compact subsets of E and $C_b(X)$, resp. Furthermore, suppose that the mapping*

$$\sim \; : \; E \to \mathrm{Mod}(\mathrm{Mod}_S(E,C_b(X)) \qquad \sigma \to \tilde{\sigma} \; ; \; \tilde{\sigma}(T) = T(\sigma)$$

is bijective. If

$$b \; : \; \mathrm{Mod}_S(E,C_b(X)) \times M_{S'}(X) \to \mathbb{K}$$

is bilinear, hypocontinuous and compatible with the $C_b(X)$*-module structure, then there is a unique* $\sigma_b \in E$ *such that*

$$b(T,\mu) = \mu(T(\sigma_b)) \quad \text{for all } (T,\mu) \in \mathrm{Mod}(E,C_b(X)) \times M(X)$$

i.e. we have $b_{\sigma_b} = b$.

Proof. Fix a $T \in \mathrm{Mod}(E,C_b(X))$. Then the mapping $\mu \to b(T,\mu)$: $M_{S'}(X) \to \mathbb{K}$ is continuous and linear. Hence there is a unique $\Phi(T) \in C_b(X)$ such that $b(T,\mu) = \mu(\Phi(T))$. As b is bilinear, the mapping $\Phi : \mathrm{Mod}(E,C_b(X)) \to C_b(X)$ will be linear. Moreover, this mapping is a $C_b(X)$-module homomorphism, as the following calculation shows:

For all $\mu \in M(X)$ we have

$$\begin{aligned} \mu(\Phi(f\cdot T)) &= b(f\cdot T,\mu) \\ &= b(T,f\cdot\mu) \\ &= (f\cdot\mu)(\Phi(T)) \\ &= \mu(f\cdot\Phi(T)), \end{aligned}$$

i.e. $\Phi(f\cdot T) = f\cdot\Phi(T)$.

Further, if we equip $\mathrm{Mod}(E,C_b(X))$ with the S-topology, then Φ is continuous. Indeed, we have

$$\begin{aligned} \|\Phi(T)\| \le 1 \quad &\text{iff} \quad |\mu(\Phi(T))| \le 1 \text{ for all } \mu \text{ with } \|\mu\| \le 1 \\ &\text{iff} \quad |b(T,\mu)| \le 1 \text{ for all } \mu \text{ with } \|\mu\| \le 1 \\ &\text{iff } T \in \{T' : |b(T,\mu)| \le 1 \text{ whenever } \|\mu\| \le 1\}. \end{aligned}$$

But this last set is open by the hypocontinuity of b. As the mapping $\sim$ is surjective, there is a $\sigma_b \in E$ such that $\Phi(T) = T(\sigma_b)$ for all $T \in \mathrm{Mod}(E, C_b(X))$ and for this σ_b we have $b(T,\mu) = \mu(\Phi(T)) = \mu(T(\sigma_b))$. The uniqueness of σ_b follows easily from the injectivity of $\tilde{}$. □

21.4 Proposition. *Let* $p : E \to X$ *be a Mackey bundle and let* G *be a topological* $C_b(X)$*-module. If*

$$b : \mathrm{Mod}_{cc}(\Gamma(p), C_b(X)) \times M_c(X) \to G$$

is a hypocontinuous bilinear mapping between $C_b(X)$*-modules, then there is a unique continuous* $C_b(X)$*-module homomorphism*

$$S_b : G'_c \to \Gamma(p)$$

such that $\lambda \circ b(T,\mu) = \mu(T \circ S_b(\lambda))$ *for all* $T \in \mathrm{Mod}(\Gamma(p), C_b(X))$, *all* $\mu \in M(X)$ *and all* $\lambda \in G'$.

Proof. Let $\lambda \in G'$. Then, applying (21.3), we can find a unique $S_b(\lambda) \in \Gamma(p)$ such that $(\lambda \circ b)(T,\mu) = \mu(T \circ S_b(\lambda))$. Obviously, the mapping $S_b : G' \to \Gamma(p)$ is linear. Moreover, given $\mu \in M(X)$, $T \in \mathrm{Mod}(\Gamma(p), C_b(X))$ $f \in C_b(X)$ and $\lambda \in G'$, we compute

$$\begin{aligned}(T(S_b(f\cdot\lambda)) &= (f\cdot\lambda)\circ b(T,\mu) \\ &= \lambda(f\cdot b(T,\mu)) \\ &= \lambda(b(f\cdot T,\mu)) \\ &= \mu(f\cdot T(S_b(\lambda)) \\ &= \mu(T(f\cdot S_b(\lambda)).\end{aligned}$$

As this holds for all μ and all T and as $\mathrm{Mod}(\Gamma(p), C_b(X))$ separates the points of $\Gamma(p)$, we conclude that

$$S_b(f\cdot\lambda) = f\cdot S_b(\lambda) \qquad \text{for all } f \in C_b(X),\ \lambda \in G'.$$

i.e. S_b is a $C_b(X)$-module homomorphism.

It remains to show that S_b is continuous. Let $U \subset \Gamma(p)$ be an open neighborhood of O. As $p : E \to X$ is a Mackey bundle, we may assume that U is of the form

$$U \;=\; \{\sigma \in \Gamma(p) \;:\; \|T(\sigma)\| \leq 1 \text{ for all } T \in A\},$$

where $A \subset \mathrm{Mod}_{cc}(\Gamma(p),C_b(X))$ is a compact convex and circled subset. Define $B := \{\mu \in M(X) \;:\; \|\mu\| \leq 1\}$. Then the restriction of b to $A\times B$ is continuous. Thus, the image of $A\times B$ under b is compact in G and an easy calculation shows that $S_b(\lambda) \in U$ if and only if $\lambda \in \{\gamma \in G' : |\gamma\circ b(T,\mu)| \leq 1 \text{ for all } (T,\mu) \in A\times B\}$. Since the latter set is open in G'_c, this shows the continuity of S_b. □

We are now in the position to identify the dual space $\Gamma(p)'$ of $\Gamma(p)$ as a certain "tensor product" in the category of $C_b(X)$-modules:

21.5 Theorem. *Let* $p : E \to X$ *be a Mackey bundle and let* G *be a quasicomplete topological* $C_b(X)$*-module. If*

$$b \;:\; \mathrm{Mod}_{cc}(\Gamma(p),C_b(X))\times M_c(X) \;\to\; G$$

is a hypocontinuous bilinear map between $C_b(X)$*-modules, then there is a unique continuous* $C_b(X)$*-module homomorphism*

$$\tilde{b} \;:\; \Gamma(p)'_{cc} \;\to\; G$$

such that $\tilde{b}(T\boxtimes\mu) = b(T,\mu)$ *for all* $T \in \mathrm{Mod}(\Gamma(p),C_b(X))$ *and all* $\mu \in M(X)$.

Moreover, if $p : E \to X$ *is in addition a bundle of Banach spaces (more generally: if* $\Gamma(p)$ *is barreled or bornological), then* $\Gamma(p)'_{cc}$ *is uniquely determined by this property in the following sense:*

Given a quasicomplete topological $C_b(X)$*-module* T *and a hypocontinuous bilinear mapping between* $C_b(X)$*-modules*

$$\tau \;:\; \mathrm{Mod}_{cc}(\Gamma(p), C_b(X)) \times M_c(X) \;\rightarrow\; T$$

such that every hypocontinuous and bilinear mapping between $C_b(X)$*-modules* $b : \mathrm{Mod}_{cc}(\Gamma(p), C_b(X)) \times M_c(X) \rightarrow G$ *into a quasicomplete topological* $C_b(X)$*-module, is of the form* $b = \tilde{b}\circ\tau$ *for a unique continuous* $C_b(X)$*-module homomorphism* $\tilde{b} : T \rightarrow G$*, then* T *is (topologically) isomorphic to* $\Gamma(p)'_{cc}$.

Proof. Let $S_b : G'_c \rightarrow \Gamma(p)$ be the mapping constructed in (21.4) and let $\tilde{b} : \Gamma(p)' \rightarrow (G'_c)'$ be its adjoint. As G is quasicomplete, the topology of compact convergence on G' is the Mackey topology. Hence we obtain $(G'_c)' \simeq G$ and $\tilde{b}$ will be continuous for the Mackey topologies on $\Gamma(p)'$ and G resp. As the original topology on G is coarser than the Mackey topology $\tau(G,G')$, we conclude that $\tilde{b} : \Gamma(p)'_{cc} \rightarrow G$ is continuous.

Clearly, $\tilde{b}$ will be a $C_b(X)$-module homomorphism. Moreover, for all $\lambda \in G'$ we have

$$\begin{aligned} \lambda(\tilde{b}(T\otimes\mu)) &= (T\otimes\mu)(S_b(\lambda)) \\ &= \mu(T(S_b(\lambda))) \\ &= \lambda\circ b(T,\mu), \end{aligned}$$

i.e. $\tilde{b}(T\ \mu) = b(T,\mu)$.

From (21.2(iii)) we conclude that $\tilde{b}$ is uniquely determined.

The second half of (21.5) follows from general category theory, if on recalls that $\Gamma(p)'_{cc}$ will be quasicomplete whenever $\Gamma(p)$ is barreled or bornological. □

In the special case where the bundle $p : E \to X$ is trivial, the dual of $\Gamma(p)$ (and hence the tensor product over $C_b(X)$ between $M_c(X)$ and $Mod_{cc}(\Gamma(p), C_b(X))$) may be represented more explicitly (see [Gr 55], [Si 59], [We 59], [Cá 66], [Su 69] and especially [Pr 77]). We shall use a different approach in this paper which makes use of the M-structure of $\Gamma(p)$.

Before we start the final pages of these notes, I would like to remark that all the following ideas are based on a joint work together with Klaus Keimel done in 1976, which however never was published. To indicate this fact, we shall assign a new head line to these final pages:

APPENDIX

Integral Representation of Linear Functionals on Spaces of Functions

by

Gerhard Gierz and Klaus Keimel

From now on, let $p : E \to X$ be a fixed bundle of real Banach spaces over a compact base space X. We shall concern ourselves with the following problems:

Problem A. Given a continuous linear functional $\phi \in \Gamma(p)'$, find a regular Borel measure μ on X and a function $\eta : X \to \Gamma(p)'$ such that the mapping $x \to \eta(x)(\sigma)$ is μ-integrable for every $\sigma \in \Gamma(p)$ and such that

$$\phi(\sigma) = \int_X \eta(x)(\sigma) \, d\mu$$

Problem B. Choose η such that $\eta(x) \in E'_x$ and $||\eta(x)|| \leq 1$ for all $x \in X$. In this case we would have $\eta(x)(\sigma) = \eta(x)(\sigma(x))$ for all $x \in X$, i.e.

$$\phi(\sigma) = \int_X \eta(x)(\sigma(x))\, d\mu$$

and

$$\begin{aligned} T_\eta \;:\; \Gamma(p) &\to L^\infty(X,\mu) \\ \sigma &\to T_\eta(\sigma);\; T_\eta(\sigma)(x) = \eta(x)(\sigma(x)) \end{aligned}$$

is a continuous $C_b(X)$-module homomorphism, where $L^\infty(X,\mu)$ denotes the space of all bounded μ-integrable functions, equipped with the supremum norm.

Problem C. Choose η as in problem B, but try to obtain in addition that $T_\eta(\sigma)$ is Borel measurable for every $\sigma \in \Gamma(p)$.

It turns out that different technics can be used to solve these problems. We could apply a Strassen desintegration theorem in the form stated by M.Neumann in [Ne 77], we could work with vector valued martingales as it was done by M.Métivier in [Me 67] or we could use a vector-valued Radon-Nikodym theorem. As we think that the last method is the most instructive one, we shall develop a Radon-Nikodym theorem which is taylored to our problem. Here, of course, most of the work was already done by various other authors. We shall follow the ideas of J.Kupka as they were carried out in [Ku 77].

Let us start with an element $f \in C(X)$. Recall from (13.18) and (13.19) that the operator

$$\begin{aligned} \hat{f} \;:\; \Gamma(p)' &\to \Gamma(p)' \\ \phi &\to f\cdot\phi \;;\; (f\cdot\phi)(\sigma) = \phi(f\cdot\sigma) \end{aligned}$$

belongs to the Cunningham algebra $Cu(\Gamma(p)')$ of $\Gamma(p)'$ and that

$$\begin{aligned} \hat{} : C(X) &\to Cu(\Gamma(p)') \\ f &\to \hat{f} \end{aligned}$$

is a norm preserving mapping between Banach algebras (Banach lattices resp.).

If $A \subset X$ is closed in X, then $N_A = \{\sigma \in \Gamma(p) : \sigma_{/A} = 0\}$ is an M-ideal of $\Gamma(p)$ by (13.6) and thus there is an L-projection

$$p_A : \Gamma(p)' \to N_A^O \simeq \Gamma_A(p)'.$$

21.6 Proposition. $p_A = \inf \{\hat{f} : f \in C(X),\ 0 \le f \le 1 \text{ and } f_{/A} = 1\}$
$= \inf \{\hat{f} : f \in C(X),\ 0 \le f \le 1 \text{ and } f_{/U} = 1$ for some open neighborhood $U \supset A\}$.

Proof. Let

$$I_A = \{\hat{f} : f \in C(X),\ 0 \le f \le 1 \text{ and } f_{/A} = 1\}$$

and

$$\tilde{I}_A = \{\hat{f} : f \in C(X),\ 0 \le f \le 1 \text{ and } f_{/A} = 1 \text{ where } U \text{ is open, } A \subset U\}.$$

As I_A and $\tilde{I}_A$ are closed under multiplication, we may use (13.2) to see that $\inf I_A$ and $\inf \tilde{I}_A$ exist in $Cu(\Gamma(p)')$ and are idempotent. Moreover, $I_A \supset \tilde{I}_A$ implies $\inf I_A \le \inf \tilde{I}_A$.
Let $\hat{f} \in I_A$, $\phi \in \Gamma(p)'$ and $\sigma \in \Gamma(p)$. Then $(1 - f)\cdot\sigma \in N_A$ and therefore $((id - \hat{f})\circ p_A)(\phi)(\sigma) = p_A(\phi)((1 - f)\cdot\sigma) = 0$, as $p_A(\phi) \in N_A^O$. Because σ and ϕ were arbitrary, we may conclude that $\hat{f}\circ p_A = p_A$. This yields $p_A \le \hat{f}$ in $Cu(\Gamma(p)')$ as we have $0 \le f \le 1$, i.e. $p_A \le \inf I_A$.
It remains to show that $\inf \tilde{I}_A \le p_A$. This statement is equivalent to

$(\inf \tilde{I}_A)(\phi) \in N_A^o$ for every $\phi \in \Gamma(p)'$, because this would imply $p_A \circ \inf \tilde{I}_A = \inf \tilde{I}_A$, i.e. $p_A \geq \inf \tilde{I}_A$ as $\inf \tilde{I}_A$ is idempotent.
Let $\phi \in \Gamma(p)'$, let $\sigma \in N_A$ and let $\varepsilon > 0$. Define

$$V := \{x \in X : \|\sigma(x)\| < \varepsilon\}$$

Choose an open set $U \subset X$ such that $A \subset U \subset \overline{U} \subset V$ and let $f : X \to [0,1]$ be a continuous function vanishing on the complement of V and taking the value 1 on U. Then we may conclude that $\|g \cdot \sigma\| \leq \varepsilon$ whenever $0 \leq g \leq f$, $g \in C(X)$. Therefore for all $\hat{g} \in \tilde{I}_A$ with $\hat{g} \leq \hat{f}$ we obtain the inequality

$$\begin{aligned} |\hat{g}(\phi)(\sigma)| &= |\phi(g \cdot \sigma)| \\ &\leq \|\phi\| \cdot \varepsilon \end{aligned}$$

By (13.2), the net $\{\hat{h}(\phi) : \hat{h} \in \tilde{I}_A\}$ converges to $(\inf \tilde{I}_A)(\phi)$ in the norm topology of $\Gamma(p)'$. Since norm convergence implies weak-*convergence, we have

$$(\inf \tilde{I}_A)(\phi)(\sigma) \leq \|\phi\| \cdot \varepsilon$$

and as $\varepsilon > 0$ was arbitrary, we conlcude that $(\inf \tilde{I}_A)(\phi)(\sigma) = 0$, i.e. $(\inf \tilde{I}_A)(\phi) \in N_A^o$. □

21.7 Proposition. *(i) If* $A,B \subset X$ *are closed subsets, then* $p_{A\cup B} = p_A \vee p_B$.
(ii) If $(A_i)_{i\in I}$ *is a family of closed subsets of* X *and if* $A = \bigcap_{i\in I} A_i$, *then* $p_A = \inf_{i\in I} p_{A_i}$.
In both cases, the lattice operations are taken in the Banach lattice $Cu(\Gamma(p)')$

Proof. (i) It is a well known fact from [AE 72] that the sum of two $\sigma(\Gamma(p)',\Gamma(p))$-closed L-ideals of $\Gamma(p)'$ is again $\sigma(\Gamma(p)',\Gamma(p))$-closed (the proof of this fact uses (13.4) and the Krein-Smulian theorem). Hence we have

$$N_A^o + N_B^o = (N_A \cap N_B)^o = N_{A\cup B}^o$$

and therefore $p_A \vee p_B = p_{A\cup B}$.

(ii): We conclude from (15.7(ii)) that

$$N_A^o = \bigcap_{i\in I} N_{A_i}^o,$$

i.e. $p_A = \inf_{i\in I} p_{A_i}$. □

We now extend the mapping $A \to p_A : Cl(X) \to Cu(\Gamma(p)')$ to all Borel subsets of X:

21.7 Definition. (i) If $U \subset X$ is an open subset of X, then we define $p_U := id - p_{X\setminus U}$.

(ii) For every subset $M \subset X$ we let

$$p_*(M) = \sup\{p_A : A = \bar{A} \subset M\}$$
$$p^*(M) = \inf\{p_U : M \subset U,\ U \text{ open}\}.$$

(iii) Let $B_p(X) := \{M \in X : p_*(M) = p^*(M)\}$. If $M \subset B_p$, then we define $p_M := p_*(M) = p^*(M)$. □

21.8 Proposition. *(i) The mappings* $p_*, p^* : 2^X \to Cu(\Gamma(p)')$ *are monotone and* $p_* \le p^*$.

(ii) $p_*(X \setminus M) + p^*(M) = id$.

(iii) A subset $M \subset X$ *belongs to* $B_p(X)$ *if and only if for every* $\phi \in \Gamma(p)'$ *and every* $\varepsilon > 0$ *there are a closed set* $A \subset M$ *and an open set* $U \supset M$ *such that* $\|p_U(\phi) - p_A(\phi)\| < \varepsilon$.

Proof. (i) Obviously, the mappings p^* and p_* are monotone. Moreover, if $M \subset X$ is given and if $A \subset M$ is closed and if $U \supset M$ is open, then $A \cap (X \setminus U) = \emptyset$ and hence $p_A \wedge p_{X\setminus U} = 0$ by (21.6). This implies the inequality $p_A \le id - p_{X\setminus U} = p_U$. As A and U were arbitrary, this yields $p_*(M) \le p^*(M)$.

(ii) follows from the computation

$$\begin{aligned} p_*(X \setminus M) &= \sup \{p_A : A = \overline{A} \subset X \setminus M\} \\ &= \sup \{p_{X\setminus U} : M \subset U, U \text{ open}\} \\ &= \sup \{id - p_U : M \subset U, U \text{ open}\} \\ &= id - \inf \{p_U : M \subset U, U \text{ open}\} \\ &= id - p^*(U). \end{aligned}$$

(iii) follows from the fact that for every $\phi \in \Gamma(p)'$ the net $\{p_A(\phi) : A = \overline{A} \in M\}$ (resp. $\{p_U(\phi) : M \in U, U \text{ open}\}$) converges to $p_*(M)(\phi)$ (resp. $p^*(M)(\phi)$) in the norm topology (see (13.2)).

21.9 Proposition. *Let* M_n, $n \in \mathbb{N}$, *be a countable family of subsets of* X. *Then we have*

(i) $p_*(\bigcap_{n\in\mathbb{N}} M_n) = \inf_{n\in\mathbb{N}} p_*(M_n)$

(ii) $p^*(\bigcup_{n\in\mathbb{N}} M_n) = \sup_{n\in\mathbb{N}} p^*(M_n)$

Proof. (i) Every vector lattice satisfies the equation $\sup D \wedge \sup E = \sup \{d \wedge e : d \in D, e \in E\}$ whenever these suprema exist (see [Sch 77]). For given subsets $M,N \subset X$, this implies the equation

$$\begin{aligned} p_*(M) \wedge p_*(N) &= \sup \{p_A : A = \overline{A} \subset M\} \wedge \sup \{p_B : B = \overline{B} \subset N\} \\ &= \sup \{p_A \wedge p_B : A = \overline{A} \subset M, B = \overline{B} \subset N\} \\ &= \sup \{p_{A\cap B} : A = \overline{A} \subset M, B = \overline{B} \subset N\} \\ &= \sup \{p_A : A = \overline{A} \subset M \cap N\} \end{aligned}$$

$$= p_*(M \cap N).$$

Therefore, we may assume that $M_{n+1} \subset M_n$ for all $n \in \mathbb{N}$.

Now let $\phi \in \Gamma(p)'$. We have to show that $p_*(\cap M_n)(\phi) = (\inf p_*(M_n))(\phi)$. Let $\varepsilon > 0$. By (13.2) the net $\{p_A(\phi) : A = \overline{A} \subset \cap M_n\}$ converges to $p_*(\cap M_n)(\phi)$ in the norm topology of $\Gamma(p)'$. Fix a closed subset $A \subset M_n$ such that

$$\|p_B(\phi) - p_*(\cap M_n)(\phi)\| \leq \varepsilon \qquad \text{whenever } A \subset B = \overline{B} \subset \cap M_n.$$

Moreover, using (13.2) again, for every $n \in \mathbb{N}$ there is a closed subset $A_n \subset X$ such that $A \subset A_n \subset M_n$ and such that

$$\|p_{B_n}(\phi) - p_*(M_n)(\phi)\| < (\tfrac{1}{2})^n \cdot \varepsilon.$$

Let $B_n := \bigcap_{i=1}^{n} A_i$. We claim that for all $n \in \mathbb{N}$ we have

$$\|p_{B_n}(\phi) - p\ (M_n)(\phi)\| < \sum_{i=1}^{n} (\tfrac{1}{2})^i \cdot \varepsilon.$$

This inequality is obvious for $n = 1$. Suppose that the inequality holds for $n \in \mathbb{N}$. As $B_{n+1} = A_{n+1} \cap B_n$ and as $M_{n+1} \subset M_n$, we conclude that $p_{A_{n+1}} \wedge p_{B_n} = p_{A_{n+1}} \circ p_{B_n} = p_{B_{n+1}}$ and $p_*(M_{n+1}) \circ p_*(M_n) = p_*(M_{n+1})$. Since the Cunningham algebra $Cu(\ (p)')$ is commutative, this yields

$$\begin{aligned}
\|p_{B_{n+1}}(\phi) - p_*(M_{n+1})(\phi)\| &= \|p_{A_{n+1}} \circ p_{B_n}(\phi) - p_*(M_{n+1})(\phi)\| \\
&= \|p_{B_n} \circ p_{A_{n+1}}(\phi) - p_{B_n} \circ p_*(M_{n+1})(\phi) + \\
&\qquad + p_*(M_{n+1}) \circ p_{B_n}(\phi) - p_*(M_{n+1})(\phi)\| \\
&\leq \|p_{B_n}\| \cdot \|p_{A_{n+1}}(\phi) - p_*(M_{n+1})(\phi)\| + \\
&\qquad + \|p_*(M_{n+1}) \circ (p_{B_n}(\phi) - p_*(M_n)(\phi))\|
\end{aligned}$$

$$\leq \left(\tfrac{1}{2}\right)^{n+1}\cdot\varepsilon + \|p_*(M_{n+1})\|\cdot\|p_{B_n}(\phi) - p_*(M_n)(\phi)\|$$

$$< \left(\tfrac{1}{2}\right)^{n+1}\cdot\varepsilon + \sum_{i=1}^{n}\left(\tfrac{1}{2}\right)^{i}\cdot\varepsilon$$

and our inequality holds for n+1, too.

In particular, we have

$$(\ast)\qquad \|p_{B_n}(\phi) - p_*(M_n)(\phi)\| < \varepsilon \qquad \text{for all } n \in \mathbb{N}.$$

Now let $B = \bigcap_{n\in\mathbb{N}} B_n$. Then $A \subset B \subset \cap M_n$ and thus

$$\|p_B(\phi) - p_*(\cap M_n)(\phi)\| < \varepsilon.$$

As $p_B(\phi) = \lim_{n\to\infty} p_{B_n}(\phi)$ and as $\inf_{n\in\mathbb{N}} p_*(M_n)(\phi) = \lim_{n\to\infty} p_*(M_n)(\phi)$ by (9.2), we conclude from $(\ast)$ that

$$\|p_B(\phi) - \inf_{n\in\mathbb{N}} p_*(M_n)(\phi)\| \leq \varepsilon$$

and the triangle inequality yields

$$\|\inf_{n\in\mathbb{N}} p_*(M_n)(\phi) - p_*(\cap M_n)(\phi)\| \leq 2\cdot\varepsilon$$

As $\varepsilon > 0$ was arbitrary, we obtain $\inf_{n\in\mathbb{N}} p_*(M_n) = p_*(\cap M_n)$.

(ii) follows from (i) and (21.8(ii)). □

21.10 Proposition. *If* $A \subset X$ *is closed, then* $p_A = p_*(A) = p^*(A)$. *In particular, we have* $A \subset B_p(X)$.

Proof. It follows from the definition of p_* that we have $p_A = p_*(A)$.

Next, let $U \subset X$ be open, let $B \subset X$ be closed and assume that $U \subset B$. Then we have $p_B \leq p_U$: Indeed, $B \cup (X \setminus U) = X$ implies $p_B \vee p_{X\setminus U} = \mathrm{id}$.

As $p_B \vee p_{X\setminus U} = p_B + p_{X\setminus U} - p_B \wedge p_{X\setminus U}$, this implies $p_B + p_{X\setminus U} \geq \mathrm{id}$, i.e. $p_B \geq \mathrm{id} - p_{X\setminus U} = p_U$.

As X is compact, we have $\cap \{\overline{U} : A \subset U,\ U \text{ open}\} = A$. Therefore we may use (21.7(ii)) to calculate

$$\begin{aligned} p^*(A) &= \inf \ \{p_U : A \subset U,\ U \text{ open}\} \\ &\leq \inf \ \{p_{\overline{U}} : A \subset U,\ U \text{ open}\} \\ &= p_A \ . \end{aligned}$$

We always have $p_A = p_*(A) \leq p^*(A)$, hence $p_A = p^*(A)$. □

21.11 Proposition. *If $p : E \to X$ is a bundle of Banach spaces over a compact base space, then $B_p(X)$ is a σ-complete Boolean algebra containing all Borel subsets of X. Moreover, the mapping*

$$\begin{aligned} p_- \ : \ B_p(X) &\to Cu(\Gamma(p)') \\ M &\to p_M \end{aligned}$$

is a σ-homomorphism between $B_p(X)$ and the (complete) Boolean algebra of all L-projections of $\Gamma(p)'$.

Proof. From (21.9) we know that $M \in B_p(X)$ if and only if $X \setminus M \in B_p(X)$ and from (21.10) we conclude that $B_p(X)$ is σ-complete and that p_- is a σ-homomorphism. Finally, all Borel subsets are contained in $B_p(X)$ by (21.10). □

Now let $\phi \in \Gamma(p)'$ be a continuous linear functional on $\Gamma(p)$. For every $M \in B_p(X)$ we define

$$\begin{aligned} \nu_\phi(M) &:= p_M(\phi) \\ \mu_\phi(M) &:= \|p_M(\phi)\| \ . \end{aligned}$$

21.12 Proposition. *If* M_n, $n \in \mathbb{N}$, *is a pairwise disjoint family of elements of* $B_p(X)$, *then*

$$\nu_\phi\left(\bigcup_{n\in\mathbb{N}} M_n\right) = \sum_{n=1}^{\infty} \nu_\phi(M_n)$$

for every $\phi \in \Gamma(p)'$. *Here the sum converges in the norm topology of* $\Gamma(p)'$ *and we have*

$$\mu_\phi\left(\bigcup_{n\in\mathbb{N}} M_n\right) = \left\|\nu_\phi\left(\bigcup_{n\in\mathbb{N}} M_n\right)\right\| = \sum_{n=1}^{\infty} \|\nu_\phi(M_n)\| = \sum_{n=1}^{\infty} \mu_\phi(M_n).$$

In particular, the mapping ν_ϕ *is a* σ*-additive* $\Gamma(p)'$*-valued measure and* μ_ϕ *is a* σ*-additive real-valued measure on* $B_p(X)$.
Finally, we have $\nu_\phi(M) = 0$ *if and only if* $\mu_\phi(M) = 0$.

Proof. Firstly, let $M,N \in B_p(X)$ be two disjoint sets. Then $p_M \wedge p_N = 0$, and whence $p_M \le id - p_N$. This implies $p_M(\phi) \in (id - p_N)(\Gamma(p)')$ and $p_N(\phi) \in p_N(\Gamma(p)')$. As p_N is an L-projection, we conclude $\|p_M(\phi) + p_N(\phi)\| = \|p_M(\phi)\| + \|p_N(\phi)\|$.
Thus, if M_n, $n \in \mathbb{N}$, is a pairwise disjoint family of elements of $B_p(X)$, we use (13.2) to calculate

$$\begin{aligned}
\nu_\phi(\cup M_n) &= p_{\cup M_n}(\phi) \\
&= (\sup_{n\in\mathbb{N}} p_{\bigcup_{i\le n} M_i})(\phi) \\
&= \lim_{n\to\infty} (p_{\bigcup_{i\le n} M_i}(\phi)) \\
&= \lim_{n\to\infty} \left(\sum_{i=1}^{n} p_{M_i}(\phi)\right) \\
&= \sum_{i=1}^{\infty} p_{M_i}(\phi)
\end{aligned}$$

and

$$\begin{aligned}
\mu_\phi(\cup M_n) &= \|\nu_\phi(\cup M_n)\| \\
&= \left\|\lim_{n\to\infty} \sum_{i=1}^{n} p_{M_i}(\phi)\right\|
\end{aligned}$$

$$= \lim_{n\to\infty} \| \sum_{i=1}^{n} p_{M_i}(\phi) \|$$

$$= \lim_{n\to\infty} \sum_{i=1}^{n} \|p_{M_i}(\phi)\|$$

$$= \sum_{i=1}^{\infty} \|p_{M_i}(\phi)\|$$

$$= \sum_{i=1}^{\infty} \mu_\phi(M_i). \quad \square$$

We are now in the position to solve problem A: Joseph Kupka has shown in [Ku 72, 4.9] that there is a function $\eta_\phi : X \to \Gamma(p)'$ such that

$$\nu_\phi(M) = \int_M \eta_\phi \cdot d\mu_\phi \qquad \text{for all } M \in B_p(X)$$

meaning that

$$p_M(\phi)(\sigma) = \int_M \eta_\phi(x)(\sigma)\cdot d\mu_\phi(x) \qquad \text{for all } M \in B_p(X),\ \sigma \in \Gamma(p).$$

Since $p_X(\phi) = \phi$, we obtain

$$\phi(\sigma) = \int_X \eta_\phi(x)(\sigma)\cdot d\mu_\phi(x) \qquad \text{for all } \sigma \in \Gamma(p),$$

as desired.

We should remark at this point that (4.9.2) of [Ku 72] provides us with more information, namely

$$\mu_\phi(M) = \|\nu_\phi(M)\|$$
$$= \int_X \|\eta_\phi(x)\| \cdot d\mu_\phi(x) \qquad \text{for all } M \in B_p(X)$$

and hence $\|\eta_\phi(x)\| = 1$ μ-almost everywhere.

In order to solve problems B and C, we need more information on the natur of η. To obtain this extra information, we find it convenient to repeat the steps of the proof of J. Kupka's result.

21.13 Definition. (i) Let (X,S,μ) be a finite measure space. By

$M^\infty(\mu)$ we denote the space of all bounded μ-measurable and real-valued functions on X.

(ii) A mapping $\rho : M^\infty(\mu) \to M^\infty(\mu)$ is called a *lifting*, provided that

(α) ρ is linear, positive and preserves multiplication and the constant function 1.

(β) $\rho(f) = f$ μ-a.e.

(γ) $f = g$ μ-a.e. implies $\rho(f) = \rho(g)$.

(iii) If, in addition, X is a compact space and if μ is a Borel measure on X, then a lifting ρ is called *almost strong*, provided that there is a subset $N \subset X$ with $\mu(N) = 0$ and

$$\rho(f)(x) = f(x) \quad \text{for all } x \in X \setminus N \text{ and all } f \in C(X).$$

21.14 Remark. Let X be a compact space, let μ be a regular Borel measure on X and let ρ be an almost strong lifting. Then there is a lifting $\rho' : M^\infty(\mu) \to M^\infty(\mu)$ such that

$$(*) \quad \rho'(f)(x) = f(x) \quad \text{for all } x \in \text{supp}(\mu) \text{ and all } f \in C(X).$$

and every lifting satisfying (*) is almost strong.

(Indeed, if μ is a regular Borel measure, then $\mu(X \setminus \text{supp}(\mu)) = 0$; hence

Conversely, let $N \subset X$ be a μ-zero set such that every continuous function f agrees with $\rho(f)$ on $X \setminus N$ and let $M = N \cap \text{supp}(\mu)$. Introduce a seminorm $||\cdot||$ on $M^\infty(\mu)$ by defining

$$||f|| = \inf \{M \in \mathbb{R} : \mu\{x \in X : |f(x)| > M\} = 0\}$$

(i.e. $||f||$ is the essential supremum of f) and let

$$N^\infty = \{f \in M^\infty(\mu) : ||f|| = 0\}.$$

Furthermore, let $\tilde{} : f \to \tilde{f} : M^\infty(\mu) \to M^\infty(\mu)/N^\infty$ be the quotient map.

Then the mapping $\tilde{f} \to \|f\|$ is well-defined and a norm on $M^\infty(\mu)/N^\infty$ and $M^\infty(\mu)/N^\infty$ is a Banach algebra in this norm. Moreover, if $f \in C(X)$, then $\tilde{f} = 0$ if and only if $f_{/\mathrm{supp}(\mu)} = 0$ and we have

$$\|\tilde{f}\| = \sup \{|f(x)| : x \in \mathrm{supp}(\mu)\}.$$

Especially, the image of $C(X)$ under $\tilde{}$ is isometrically isomorphic to $C(\mathrm{supp}(\mu))$ and therefore is closed.
For every $t \in M = N \cap \mathrm{supp}(\mu)$ we define

$$\varepsilon_t : \{\tilde{f} : f \in C(X)\} \to \mathbb{R}$$
$$\tilde{f} \to f(t).$$

Then ε_t is well-defined and continuous. Thus, using [IT 69, VIII.1, Prop. 1], we may find a continuous extension

$$\chi_t : M^\infty(\mu)/N^\infty \to \mathbb{R}$$

of ε_t such that

$$\chi_t(\tilde{f}\cdot\tilde{g}) = \chi_t(\tilde{f})\cdot\chi_t(\tilde{g}) \qquad \text{for all } \tilde{f},\tilde{g} \in M^\infty(\mu).$$

We now may define a new lifting $\rho' : M^\infty(\mu) \to M^\infty(\mu)$ by

$$\rho'(f)(x) = \begin{cases} \rho(f)(x) & \text{if } x \in X \setminus M, \\ \chi_x(\tilde{f}) & \text{if } x \in M. \end{cases}$$)

21.15 Definition. Let X be a compact space and let μ be a regular Borel measure on X. If ρ is a lifting satisfying (*) of (21.14), then ρ is called a *strong lifting*. □

Now let $\rho : M^\infty(\mu) \to M^\infty(\mu)$ be a lifting, where (X,S,μ) is a measure space. Let $\tilde{S} = \{A \in X : \chi_A \in M^\infty(\mu)\}$ be the μ-completion of S, where χ_A denotes the characteristic function of A. If χ_A belongs to $M^\infty(\mu)$, and therefore $\rho(\chi_A)$ is idempotent, too. Thus, ρ defines an element $B \in \tilde{S}$ such that $\rho(\chi_A) = \chi_B$; in this case we write $\rho^*(A) = B$.

21.16 Proposition. *(i) Let* (X,S,μ) *be a measure space and let* $\rho : M^\infty(\mu) \to M^\infty(\mu)$ *be a lifting. Then the mapping*

$$\rho^* \ : \ \tilde{S} \to \tilde{S}$$
$$\rho(\chi_A) = \chi_{\rho^*(A)}$$

is a homomorphism of Boolean algebras satisfying

β^*) $\rho^*(A) = A$ μ-a.e.

γ^*) *If* $A = B$ μ-a.e., *then* $\rho^*(A) = \rho^*(B)$.

(ii) If X *is a compact space, if* μ *is a regular Borel measure on* X *and if* $\rho : M^\infty(\mu) \to M^\infty(\mu)$ *is a strong lifting, then we have in addition*

I) $\rho^*(A) \cap \mathrm{supp}(\mu) \subset A$ *whenever* A *is closed*

II) If $x \in \mathrm{supp}(\mu)$ *and if* U *is a neighborhood of* x, *then* $x \in \rho^*(U)$.

Proof. For a proof of (i) we refer to [IT 69, III.1].
Although (ii) is certainly well-known, too, we indicate a proof: Let $A \subset X$ be closed and let U be an open neighborhood of A. Choose a continuous function $f : X \to [0,1]$ such that $f(A) = \{1\}$ and $f(X \setminus U) = \{0\}$. Then we have $\chi_A \le f$ and therefore $\chi_{\rho^*(A)} = \rho(\chi_A) \le \rho(f)$. This implies $\chi_{\rho^*(A)\cap\mathrm{supp}(\mu)} = \chi_{\rho^*(A)} \cdot \chi_{\mathrm{supp}(\mu)} \le \rho(f)\cdot\chi_{\mathrm{supp}(\mu)}$. As $\rho(f)$ and f agree on $\mathrm{supp}(\mu)$, we obtain $\chi_{\rho^*(A)\cap\mathrm{supp}(\mu)} \le f$ and as the open set U can be made arbitrarily small, this proves (I).
Property (II) now follows immediatly from (I) and De Morgan's rule. □

It may be shown that conversely properties (I) and (II) characterize strong liftings.

The following proposition ensures the existence of liftings:

21.17 Proposition. *(i) Let* X *be a compact space and let* μ *be a finite regular Borel measure on* X. *Then there is a lifting*
$\rho : M^{\infty}(\mu) \to M^{\infty}(\mu)$.
(ii) Moreover, if X *is metrizable, then there is a strong lifting*
$\rho : M^{\infty}(\mu) \to M^{\infty}(\mu)$.

For a proof see [IT 69, IV.2, theorem 3] and [IT 69, VIII.4, theorem 8]. □

It is well known that liftings may be used in the proof of the Radon--Nikodym theorem (see [Di 51a] or [IT 69]). Let us repeat some of the arguments here:

Let (X,S,μ) be a finite measure space, assume that μ is positive and let ρ be a lifting. Moreover, let λ be a second measure on (X,S) which is μ-continuous and which has the property

$$\text{(B)} \qquad \left\{\frac{\lambda(E)}{\mu(E)} : E \in S,\ \mu(E) \neq 0\right\} \quad \text{is bounded in } \mathbb{R}.$$

From the Radon-Nikodym theorem we know that there is a μ-integrable function η such that

$$\lambda(E) = \int_E \eta \cdot d\mu \qquad \text{for all } E \in S.$$

By the mean value theorem and (B) we may find a constant $M > 0$ such that $|\eta(x)| \leq M$ μ-a.e. Hence we may assume without loss of generality that $|\eta(x)| \leq M$ for all $x \in X$. In this case $\rho(\eta)$ is defined and we may assume that $\eta = \rho(\eta)$.

We now consider the set Π of all partitions $\pi = \{F_1,\ldots,F_n\}$ of X satisfying

$$F_i = \rho^*(F_i) \neq \emptyset \qquad (i = 1,\ldots,n).$$

The set Π is directed under refinement. For every $\pi \in \Pi$ we let

$$\eta_\pi = \sum_{F\in\pi} \frac{\lambda(F)}{\mu(F)}\cdot\chi_F.$$

21.18 Proposition. *The net* $(\eta_\pi)_{\pi\in\Pi}$ *converges to* η *uniformly on* X.

Proof. Let $\varepsilon > 0$ and define

$$A_n = \{x \in X : n\cdot\varepsilon \le \eta(x) < (n+1)\cdot\varepsilon\}, \qquad n \in \mathbb{Z}.$$

Then A_n belongs to S and it is easy to see that $\rho(\eta) = \eta$ implies $\rho^*(A_n) = A_n$. Let $\pi = \{A_n : A_n \neq \emptyset\}$. Then π is a partition of X belonging to Π. Now let us take any refinement $\pi' = \{F_1,\ldots,F_m\} \in \Pi$ of π and let $x \in X$ be arbitrary. Then we can find an index $i \in \{1,\ldots,m\}$ and an integer $n \in \mathbb{Z}$ such that $x \in F_i \subset A_n$.
We compute

$$\begin{aligned}|\eta_{\pi'}(x) - \eta(x)| &= \left|\frac{\lambda(F_i)}{\mu(F_i)} - \eta(x)\right| \\ &= \frac{1}{\mu(F_i)}\cdot\left|\int_{F_i} (\eta(t) - \eta(x))\cdot d\mu(t)\right| \\ &\le \frac{1}{\mu(F_i)} \int_{F_i} |\eta(t) - \eta(x)|\cdot d\mu(t) \\ &\le \frac{\mu(F_i)}{\mu(F_i)}\cdot\varepsilon \\ &= \varepsilon.\end{aligned}$$

Since $x \in X$ was arbitrary, we obtain

$$\sup_{x\in X} |\eta_{\pi'}(x) - \eta(x)| \le \varepsilon \quad \text{for all refinements } \pi' \text{ of } \pi.$$

□

Let us return to our bundle $p : E \to X$ and our continuous linear functional $\phi : \Gamma(p) \to \mathbb{R}$.

The σ-algebra $B_p(X)$ contains the σ-algebra of all Borel parts $B(X)$ and we have a regular Borel measure

$$\mu_\phi \; : \; B(X) \to \mathbb{R}$$
$$M \to \|p_M(\phi)\|$$

and a $\Gamma(p)'$-valued measure

$$\nu_\phi \; : \; B(X) \to \Gamma(p)'$$
$$M \to p_M(\phi)$$

on $B(X)$. Moreover, if $\sigma \in \Gamma(p)$ is a section, then we may define a Borel measure λ_σ on $B(X)$ by

$$\lambda_\sigma \; : \; B(X) \to \mathbb{R}$$
$$M \to p_M(\phi)(\sigma) = \nu_\phi(M)(\sigma).$$

It is obvious that λ_σ is μ-continuous for every $\sigma \in \Gamma(p)$ and that ν_ϕ, μ_ϕ and the λ_σ may be extended to the μ_ϕ-completion $B(X)^\sim$ of $B(X)$. Further, the set

$$\{\frac{\lambda_\sigma(E)}{\mu_\phi(E)} : E \in B(X)^\sim, \mu_\phi(E) \neq 0\}$$

is bounded by $\|\sigma\|$. By (21.17) we may choose a lifting $\rho : M^\infty(\mu) \to M^\infty(\mu)$. As above, let Π be the directed set of all partitions $\pi = \{F_1, \ldots, F_n\}$ of X such that $\rho(F_i) = F_i \neq \emptyset$, $i = 1, \ldots, n$.
We define

$$\eta_\pi \; : \; X \to \Gamma(p)'$$
$$\eta_\phi = \sum_{A \in \pi} \chi_A \cdot \frac{\nu_\phi(A)}{\mu_\phi(A)}$$

and for every $\sigma \in \Gamma(p)$ we define

$$\eta_{\pi,\sigma} \;:\; X \to \mathbb{R}$$
$$\eta_{\pi,\sigma} \;=\; \sum_{A\in\pi} \frac{\lambda_\sigma(A)}{\mu_\phi(A)}\cdot\chi_A$$

Obviously, we have $\eta_{\pi,\sigma}(x) = \eta_\pi(x)(\sigma)$ for all $x \in X$, $\sigma \in \Gamma(p)$. Moreover, (21.18) shows that the net $(\eta_{\pi,\sigma})_{\pi\in\Pi}$ converges uniformly to a function $\eta_\sigma \in M^\infty(\mu)$ with $\eta_\sigma = \rho(\eta_\sigma)$ such that

$$\lambda_\sigma(E) \;=\; \int_E \eta_\sigma\cdot d\mu_\phi.$$

Hence the net $(\eta_\pi)_{\pi\in\Pi}$ is a $\sigma(\Gamma(p)',\Gamma(p))$-Cauchy net. As we have $\|\eta_\pi(x)\| \le 1$ for all $x \in X$ and all $\pi \in \Pi$, the net $(\eta_\pi)_{\pi\in\Pi}$ converges uniformly towards a function $\eta_\phi : X \to \Gamma(p)'$ with $\eta_\phi(x)(\sigma) = \eta_\sigma(x)$ for all $x \in X$ and all $\sigma \in \Gamma(p)$.

Taking all these pieces of information together, we obtain

$$p_M(\phi)(\sigma) = \lambda_\sigma(M) = \int_M \eta_\sigma(x)\cdot d\mu_\phi(x) = \int_M \eta_\phi(x)(\sigma)d\mu_\phi(x).$$

If we let $M = X$, the we obtain again the solution of problem A.

21.19 Proposition. *For a given* $\sigma \in \Gamma(p)'$ *and a given lifting* $\rho : M^\infty(\mu_\phi) \to M^\infty(\mu_\phi)$ *we let*

$$A_x \;:=\; \cap\,\{(M \cap \operatorname{supp}(\mu_\phi))^- : x \in M = \rho^*(M)\}$$

for all $x \in X$. *Then for all* $x \in X$ *we have* $\eta_\phi(x) \in N^o_{A_x}$.

Proof. Let $x \in X$ and let $M \subset X$ be a subset such that $x \in M = \rho^*(M)$. Then $\pi = \{M, X \setminus M\}$ is a partition of X belonging to Π. Let $\pi' = \{F_1,\dots,F_n\}$ be a refinement of π. Then there is an $i \in \{1,\dots,n\}$ such that $x \in F_i \subset M$. Let $E \in B(X)$ be a Borel part of X such that $x \in E \subset F_i$ and $\mu_\phi(E) = \mu_\phi(F_i)$. Then it follows that $\nu_\phi(E) = \nu_\phi(F_i)$ and hence

$$\eta_{\pi'}(x) \;=\; \frac{\nu_\phi(E)}{\mu_\phi(E)}$$

Moreover, as μ_ϕ is a regular Borel measure, we know that $\mu_\phi(X \cap \text{supp}(\mu_\phi)) = 0$ and thus $\nu_\phi(X \cap \text{supp}(\mu_\phi)) = 0$. This yields

$$\eta_{\pi'}(x) = \frac{\nu_\phi(E \cap \text{supp}(\mu_\phi))}{\mu_\phi(E \cap \text{supp}(\mu_\phi))} .$$

Now note that $E \cap \text{supp}(\mu_\phi) \subset (M \cap \text{supp}(\mu_\phi))^-$, which gives us

$$\begin{aligned} \nu_\phi(E \cap \text{supp}(\mu_\phi)) &= p_{E\cap\text{supp}(\mu_\phi)}(\phi) \\ &= p_{(M\cap\text{supp}(\mu_\phi))^-} \circ p_{E\cap\text{supp}(\mu_\phi)}(\phi) \\ &\in p_{(M\cap\text{supp}(\mu_\phi))^-}(\Gamma(p)') \\ &= N^o_{(M\cap\text{supp}(\mu_\phi))^-} \end{aligned}$$

and therefore

$$\eta_{\pi'}(x) \in N^o_{(M\cap\text{supp}(\mu_\phi))^-} .$$

As π' was an arbitrary refinement of π and as $N^o_{(M\cap\text{supp}(\mu_\phi)^-}$ is $\sigma(\Gamma(p)',\Gamma(p))$-closed, we obtain

$$\eta_\phi(x) = \lim_{\pi\in\Pi} \eta_\pi(x) \in N^o_{(M\cap\text{supp}(\mu_\phi))^-} .$$

Finally, the mapping $A \to N^o_A$, $A \in Cl(X)$, preserves arbitrary intersections by (15.7(ii)) yielding that

$$\eta_\phi(x) \in N^o_{A_x} . \qquad \square$$

We now come to a solution of problem B:

21.20 Theorem. *Let* $p : E \to X$ *be a bundle of Banach spaces over a compact base space* X *and assume that every finite, regular Borel measure on* X *admits a strong lifting. (This is in particular the case if* X *is metrizable). If* $\phi : \Gamma(p) \to \mathbb{R}$ *is a continuous linear functional on* $\Gamma(p)$, *then we can find a family* $\eta_{\phi,x} \in E_x'$, $x \in X$, *and a finite regular Borel measure* μ_ϕ *on* X *such that*

i) $\|\eta_{\phi,x}\| \le 1$ *for all* $x \in X$.

ii) *The mapping* $x \to \eta_{\phi,x}(\sigma(x)) : X \to \mathbb{R}$ *is* μ_ϕ*-integrable for every* $\sigma \in \Gamma(p)$.

iii) *For every Borel set* $M \subset X$ *we have*

$$p_M(\phi)(\sigma) = \int_M \eta_{\phi,x}(\sigma(x)) \cdot d\mu_\phi(x),$$

in particular

$$\phi(\sigma) = \int_X \eta_{\phi,x}(\sigma(x)) \cdot d\mu_\phi(x).$$

Proof. Let $\sigma \in \Gamma(p)'$ and as before, let μ_ϕ be the measure defined by $\mu_\phi(M) = \|p_M(\phi)\|$. Further, let $\rho : M^\infty(\mu_\phi) \to M^\infty(\mu_\phi)$ be a strong lifting. If the mapping $\eta_\phi : X \to \Gamma(p)'$ is constructed as above, then we conclude from (21.19) that $\eta_\phi(x) \in N^o_{A_x}$ for every $x \in X$. We show:

$$A_x = \{x\} \quad \text{for every } x \in \operatorname{supp}(\mu_\phi).$$

Indeed, let A be a closed neighborhood of $x \in \operatorname{supp}(\mu_\phi)$. Then (21.16(ii)) implies that

$$x \in \operatorname{supp}(\mu_\phi) \cap \rho^*(A) \subset A.$$

If we let $M = \rho^*(A)$, then we know that $\rho^*(M) = M$ and

$$x \in (\operatorname{supp}(\mu_\phi) \cap M)^- \subset A.$$

As A was arbitrary, this yields $A_x = \{x\}$ and whence $\eta_\phi(x) \in N^o_x = E'_x$ for every $x \in \operatorname{supp}(\mu_\phi)$.

Now define a family $\eta_{\phi,x} \in E'_x$, $x \in X$, by

$$\eta_{\phi,x} = \begin{cases} \eta_\phi(x) & , \quad x \in \operatorname{supp}(\mu_\phi) \\ 0 & , \quad x \in X \setminus \operatorname{supp}(\mu_\phi). \end{cases}$$

Then $\|\eta_{\phi,x}\| \leq \|\eta_\phi(x)\| \leq 1$ for all $x \in X$. Moreover, since μ_ϕ is a regular Borel measure, the set $X \setminus \operatorname{supp}(\mu_\phi)$ has measure 0 and therefore $\eta_{\phi,x} = \eta_\phi(x)$ μ_ϕ-a.e. Hence for every $\sigma \in \Gamma(p)$ we have

$$p_M(\phi)(\sigma) = \int_M \eta_\phi(x)(\sigma)\cdot d\mu_\phi(x) = \int_M \eta_{\phi,x}(\sigma(x))\cdot d\mu_\phi(x). \quad \square$$

The following theorem is a partial solution of problem C:

21.21 Theorem. *Let* $p : E \to X$ *be a separable bundle of Banach spaces over a compact base space* X *and assume that every finite regular Borel measure on* X *admits a strong lifting. If* $\phi : \Gamma(p) \to \mathbb{R}$ *is a continuous linear functional on* $\Gamma(p)$, *then we can find a family* $\xi_{\phi,x} \in E'_x$, $x \in X$, *and a finite regular Borel measure* μ_ϕ *on* X *such that*

i) $\|\xi_{\phi,x}\| \leq 1$ *for all* $x \in X$.

ii) *the mapping* $x \mapsto \xi_{\phi,x}(\sigma(x)) : X \to \mathbb{R}$ *is Borel measurable and bounded for every* $\sigma \in \Gamma(p)$.

iii) *for every Borel part* $M \subset X$ *we have*

$$p_M(\phi)(\sigma) = \int_M \xi_{\phi,x}(\sigma(x))\cdot d\mu_\phi(x),$$

in particular

$$\phi(\sigma) = \int_X \xi_{\phi,x}(\sigma(x))\cdot d\mu_\phi(x).$$

Proof. Let $(\sigma_n)_{n\in\mathbb{N}}$ be a countable family of sections of $\Gamma(p)$ such that $\{\sigma_n(x) : n \in \mathbb{N}\}$ is dense in E_x for every $x \in X$. Further, let $(\eta_{\phi,x})_{x\in X}$ be a family of elements of E'_x, $x \in X$, such that the conditions i), ii) and iii) of (21.20) are satisfied. Then the mapping $x \mapsto \eta_{\phi,x}(\sigma_n(x))$ is μ_ϕ-integrable for every $n \in \mathbb{N}$. Hence we can find a Borel set $A_n \subset X$ with $\mu_\phi(A_n) = 0$ such that the mapping $x \mapsto \chi_{X\setminus A_n}(x)\cdot\eta_{\phi,x}(\sigma_n(x))$ is Borel measurable. Let $A = \bigcup_{n\in\mathbb{N}} A_n$. Then we still have $\mu_\phi(A) = 0$. Now define

$$\xi_{\phi,x} = \begin{cases} \eta_{\phi,x} & \text{if } x \in X \setminus A \\ 0 & \text{if } x \in A \end{cases}$$

With this definition the properties (i) and (iii) are satisfied. Define an operator

$$T : \Gamma(p) \to M^{\infty}(\mu_{\phi})$$
$$T(\sigma)(x) = \xi_{\phi,x}(\sigma(x)).$$

It remains to show that $T(\sigma)$ is Borel measurable for every $\sigma \in \Gamma(p)$. By the choice of the $\xi_{\phi,x}$, $x \in X$, this is clear for the σ_n, $n \in \mathbb{N}$. Moreover, the operator T is a C(X)-module homomorphism and we have $\|T(\sigma)\|_{\infty} \leq \|\sigma\|$, where $\|\cdot\|_{\infty}$ denotes the supremum norm on $M^{\infty}(\mu_{\phi})$ (recall that the elements of $M^{\infty}(\mu_{\phi})$ are bounded by definition). Therefore $T(\sigma)$ is Borel measurable for every element σ belonging to the closed C(X)-submodule generated by the set $\{\sigma_n : n \in \mathbb{N}\}$ and thus for every $\sigma \in \Gamma(p)$ by the Stone-Weierstraß theorem (4.3). □

We conclude this book with a description of the dual space of C(X,E), where E is a Banach space and where X is compact (see [Gr 55], [Si 59], [Cá 66], [We 69], [Su 69] and [Pr 77]).

21.22 Definition. Let E be a Banach space and let X be a compact topological space. A linear operator $u : C(X) \to E'$ is called *dominated*, if there is a positive finite Borel measure μ on X such that $\|u(f)\| \leq \int_X |f| \cdot d\mu$ for all $f \in C(X)$. □

Now let $\phi : C(X,E) \to \mathbb{R}$ be a continuous linear functional. We define an operator $u_{\phi} : C(X) \to E'$ by

(1) $\quad u_{\phi}(f)(a) = \phi(f \cdot c_a) \qquad$ for all $a \in E$

where $c_a : X \to E$ denotes the constant mapping with value a. As on the previous pages, let μ_{ϕ} be the finite Borel measure on X defined by $\mu_\phi(M) = \|p_M(\phi)\|$. We claim that

$$\|\mu_\phi(f)\| \le \int_X |f|\cdot d\mu_\phi \qquad \text{for all } f \in C(X).$$

Indeed, let $\varepsilon > 0$, let $f \in C(X)$ and define

$$A_n := \{x \in X : n\cdot\varepsilon \le f(x) < (n+1)\cdot\varepsilon\} \quad , n \in \mathbb{Z}.$$

Then for each compact subset $K \subset A_n$ we have

$$\begin{aligned} |p_K(\phi)(f\cdot c_a)| &\le \|p_K(\phi)\|\cdot\sup\{\|f(x)\cdot a\| : x \in K\} \\ &= \mu_\phi(K)\cdot\|a\|\cdot\sup\{|f(x)| : x \in K\} \\ &\le \mu_\phi(A_n)\cdot\|a\|\cdot\sup\{|f(x)| : x \in A_n\}. \end{aligned}$$

As the $p_K(\phi)$, $K \in A_n$, converge to $p_{A_n}(\phi)$ in the norm topology of $C(X,E)'$, we conclude that

$$|p_{A_n}(\phi)(f\cdot c_a)| \le \mu_\phi(A_n)\cdot\|a\|\cdot\sup\{|f(x)| : x \in A_n\}.$$

Note that the sets A_n, $n \in \mathbb{Z}$ are pairwise disjoint. Hence (21.12) yields

$$\begin{aligned} |\mu_\phi(f)(a)| &= |\phi(f\cdot c_a)| \\ &= |\sum_{n\in\mathbb{Z}} p_{A_n}(\phi)(f\cdot c_a)| \\ &\le \sum_{n\in\mathbb{Z}} |p_{A_n}(\phi)(f\cdot c_a)| \\ &\le \|a\|\cdot(\sum_{n\in\mathbb{Z}} \mu_\phi(A_n)\cdot\sup\{|f(x)| : x \in A_n\}) \\ &\le \|a\|\cdot(\sum_{n\in\mathbb{Z}} (\int_{A_n} |f(x)|\cdot d\mu_\phi + \varepsilon\cdot\mu_\phi(A_n))) \\ &= \|a\|\cdot(\int_X |f(x)|\, d\mu_\phi + \varepsilon\cdot\mu_\phi(X)) \end{aligned}$$

As $a \in E$ and $\varepsilon > 0$ were arbitrary, we conclude that

$$\|u_\phi(f)\| \le \int_X |f(x)|\cdot d\mu_\phi$$

and therefore u_ϕ is dominated.

Conversely, assume that $u : C(X) \to E'$ is dominated, i.e.

$$\|u(f)\| \leq \int_X |f(x)|\cdot d\mu$$

for a certain finite Borel measure μ on X. In this case, we define a linear functional ϕ_u on the $C(X)$-submodule $M \in C(X,E)$ spanned by the constant functions via the formula

$$(2) \qquad \phi_u\Big(\sum_{i=1}^{n} f_i\cdot c_{a_i}\Big) = \sum_{i=1}^{n} u(f_i)(a_i),$$

where the f_i belong to $C(X)$ and where the a_i belong to E_i. It is not too difficult to check that ϕ_u is well defined and linear. Moreover, if all the f_i are positive, if $\sum_{i=1}^{n} f_i = 1$ and if $\|a_i\| \leq 1$ for all $i \in \{1,\ldots,n\}$, then

$$\begin{aligned} \Big|\phi_u\Big(\sum_{i=1}^{n} f_i\, c_{a_i}\Big)\Big| &\leq \sum_{i=1}^{n} |u(f_i)(a_i)| \\ &\leq \sum_{i=1}^{n} \|u(f_i)\| \\ &\leq \sum_{i=1}^{n} \int_X f_i(x)\cdot d\mu \\ &= \int_X \sum_{i=1}^{n} f_i(x)\cdot d\mu \\ &= \|\mu\|. \end{aligned}$$

Repeating some arguments from the proof of the Stone-Weierstraß theorem, we find that ϕ_u is bounded on M, as the elements of the above form are dense in the unit ball of M. Thus, as M is dense in $C(X,E)$, the mapping ϕ_u may be uniquely extended to $C(X,E)$. Obviously, the mappings $u \to \phi_u$ and $\phi \to u_\phi$ are mutually inverse to each other. Thus, we have shown:

21.23 Theorem (Wells 1965). *Let* X *be a compact space and let* E *be a Banach space. Then the mappings defined in* (1) *and* (2) *are mutually inverse isomorphisms between the dual space* $C(X,E)'$ *of* $C(X,E)$ *and the space of all dominated operators* $u : C(X) \to E'$. □

References

[AE 72] Alfsen, E.M. and E.G.Effros: Structure in real Banach spaces I/II, Ann. of Math. 96, 98 - 173 (1972)

[Al 71] Alfsen, E.M.: Compact convex sets and boundary integrals, Springer Verlag, Berlin, Heidelberg, New York (1971)

[Au 75] Auspitz, N.E.: Dissertation, University of Waterloo (1975)

[Ba 75] Baker, C.W.: The Pedersen ideal and the representation of C^*-algebras, Dissertation, University of Kentucky, Lexingtion (1975)

[Ba 77] Banaschewski, B.: Sheaves of Banach spaces, Quaest. Math. 2, 1 - 22 (1977)

[Ba 79] Banaschewski, B.: Injective Banach sheaves, in: Applications of Sheaves, Lecture Notes in Mathematics 753, Springer Verlag, Berlin, Heidelberg, New York (1979)

[Be 79] Behrends, E.: M-structure and the Banach-Stone theorem, Lecture Notes in Mathematics 736, Springer Verlag, Berlin, Heidelberg, New York (1979)

[Bi 61] Bishop, E.: A generalization of the Stone-Weierstraß theorem, Pacific J. Math. 11, 777 - 783 (1961)

[Bi 80] Bierstedt, K.-D.: The approximation-theoretic localization of Schwartz's approximation property for weighted locally convex function spaces and some examples, to appear (1980)

[BK 41] Bohnenblust, H.F. and S.Kakutani: Concrete representation of (M)-spaces, Ann of Math. 42, 1025 - 1028 (1941)

[BM 79] Burden, C.W. and C.J.Mulvey: Banach spaces in categories of sheaves, in: Applications of Sheaves, Lecture Notes in Mathematics 753, Springer-Verlag, Berlin, Heidelberg, New York (1979)

[Bo 75] Bowshell, R.A.: Continuous associative sums of Banach spaces, Preprint (1975)

[Br 59] de Branges, L.: The Stone-Weierstraß theorem, Proc. Amer. Math. Soc. 10, 822 - 824 (1959)

[Bu 58] Buck, R.C.: Bounded continuous functions on a locally compact space, Michigan Math. J. 5, 95 - 104 (1958)

[Bu 80] Burden, C.W.: The Hahn-Banach theorem in a category of sheaves, J. Pure and Applied Algebra 17, 25 - 34 (1980)

[Cá 66] Các, N.P.: Linear transformations on some functional spaces, Proc. London Math. Soc. (3) 16, 705 - 736 (1966)

[Comp 80] Gierz, G., K.H.Hofmann, K.Keimel, J.D.Lawson, M.Mislove and D.Scott: A compendium of continuous lattices, Springer Verlag, Berlin, Heidelberg, New York (1980)

[Cr 74] Cunningham, F. Jr. and N.M.Roy: Extreme functionals on an upper semicontinuous function space, Proc. Amer. Math. Soc. 42, 461 - 465 (1974)

[Cu 67] Cunningham, F. Jr.: M-structure in Banach spaces, Proc. Cambrigde Philos. Soc. 63, 613 - 629 (1967)

[DD 63] Dixmier, J. and A.Douady: Champs contius d'espaces hilbertiens et de C^*-algèbres, Bull. Soc. Math. France 91, 227 - 284 (1963)

[DH 68] Dauns, J. and K.H.Hofmann: Representation of rings by sections, Memoirs of the Amer. Math. Soc. 83 (1968)

[DH 69] Dauns, J. and K.H.Hofmann: Spectral theory of algebras and adjunctions of identity, Math. Ann. 179, 175 - 202 (1969)

[Di 51a] Dieudonné, J.: Sur le theorème de Lebesque-Nikodym IV, J. Indian Math. Soc. 15, 77 - 86 (1951)

[Di 51b] Dieudonnê, J.: Sur le theorème de Lebesque-Nikodym V, Canad. J. Math. 3, 129 - 139 (1951)

[Du 73] Dupré, M.J.: Classifying Hilbert bundles I,II, J. of Functional Analysis, Vol 15, No 3, 244 - 322 (1973)

[El 75] Elliot, G.A.: An abstract Dauns-Hofmann-Kaplansky multiplier theorem, Canad. Math. J. 27, 827 - 836 (1975)

[EO 74] Elliot, G.A. and D. Oleson: A simple proof of the Dauns-Hofmann theorem, Math. Scand. 34, 231 - 234 (1974)

[Fe 61] Fell, J.M.G.: The structure of algebras of operator fields, Acta Math. 106, 233 - 280 (1961)

[Fe 77] Fell, J.M.G.: Induced representations and Banach-*-algebraic bundles, Lecture Notes in Mathematics 582, Springer Verlag, Berlin, Heidelberg, New York (1977)

[FGK 78] Flösser, H.O., G.Gierz and K.Keimel: Structure spaces and the ideal center of vector lattices, Quart. J. Math. Oxford (2) 29, 415 - 426 (1978)

[Gi 77] Gierz, G.: Darstellung von Banachverbänden durch Schnitte in Bündeln, Mitteilungen a. d. Mathem. Seminar Giessen, Heft 125 (1977)

[Gi 78] Gierz, G.: Representation of spaces of compact operators and applications to the approximation property, Archiv der Mathematik 30, 622 - 628 (1978)

[GK 77] Gierz, G. and K.Keimel: A lemma on primes appearing in algebra and analysis, Houston J. of Math. Vol 3, No 2, 207 - 224 (1977)

[Gl 63] Glicksberg, I.: Bishop's generalized Stone-Weierstraß theorem for the strict topology, Proc. Amer. Math. Soc. 14, 329 - 333 (1963)

[Go 49] Godement, R.: Theórie générale des sommes continues d'espaces de Banach, C.R.Acad. Sci. Paris 228, 1321 - 1323 (1949)

[Go 51] Godement, R.: Sur la théorie des représentations unitaires, Ann. of Math. 50, 68 - 124 (1951)

[Gr 55] Grothendieck, A.: Produits tensoriels topologique et espaces nucléaires, Mem. Amer. Math. Soc. 16 (1955)

[Gr 65] Gray, J.: Sheaves with values in arbitrary categories, Topology 3, 1 - 18 (1965)

[Ha 77] Haydon, R.: Injective Banach lattices, Math. Z. 156, 19 - 48 (1977)

[Hi 78] Hirzebruch, F.: Topological methods in algebraic geometry, Grundlehren der math. Wissenschaften 131, Springer--Verlag, Berlin, Heidelberg, New York (1978)

[HK 79] Hofmann, K.H. and K.Keimel: Sheaf theoretical concepts in analysis: Bundles and sheaves of Banach spaces, Banach C(X)-modules, in: Application of Sheaves, Lecture Notes in Mathematics 753, Springer-Verlag

[HL 74] Hofmann, K.H. and J. Luikkonen (eds.): Recent advances in the representation theory of rings and C^*-algebras by continuous sections, Mem. Amer. Math. Soc. 148 (1974)

[Ho 72] Hofmann, K.H.: Representation of algebras by continuous sections, Bull. Amer. Math. Soc. 78, 291 - 373 (1972)

[Ho 74] Hofmann, K.H.: Bundles of Banach spaces, sheaves of Banach spaces, C(B)-modules, Lecture Notes, TH Darmstadt (1974)

[Ho 75] Hofmann, K.H.: Sheaves and bundles of Banach spaces, Tulane University Preprint (1975)

[Ho 77] Hofmann, K.H.: Sheaves and bundles of Banach spaces are equivalent, in: Lecture Notes in Mathematics 575, 53 - 69, Springer-Verlag, Berlin, Heidelberg, New York (1977)

[Hu 66] Husemoller, D.: Fibre bundles, Mc.Graw-Hill Book Company (1966)

[IT 69] Ionescu Tulcea, A.&C.: Topics in the theory of liftings, Ergebnisse der Mathematik und ihrer Grenzgebiete 48, Springer Verlag, Berlin, Heidelberg, New York (1969)

[Ka 51] Kaplansky, I.: The structure of certain operator algebras, Trans. Amer. Math. Soc. 70, 219 - 255 (1951)

[KR 80] Kitchen, J.W. and D.A.Robbins: Tensor Products of Banach bundles, Pac. J. of Math., to appear

[Ku 72] Kupka, J.: Radon-Nikodym theorems for vector valued measures, Trans. Amer. Math. Soc. 169, 197 - 217 (1972)

[Mé 67] Métivier, M.: Martingales à valeur vectorielles, Applications à la dérivation des measures vectorielles, Annales de l'Institute Fourier 17, fasc. 2, 175 - 208 (1967)

[Mö 78] Möller, H.: Bündel von halbnormierten Räumen, Diplomarbeit, TH Darmstadt (1978)

[Mu 79] Mulvey, C.J.: Representation of rings and modules, in: Applications of Sheaves, Lecture Notes in Mathematics 753, Springer-Verlag, Berlin, Heidelberg, New York (1979)

[Mu 80] Mulvey, C.J.: Banach sheaves, J. Pure and Applied Algebra 17, 69 - 83 (1980)

[Na 59] Nachbin, L.: Algebras of finite diffential order and the operator calculus, Ann. of Math., Vol 70, No. 3, 413 - 437 (1959)

[Nai 59] Naimark, M.A.: Normierte Algebren, Dt. Verlag d. Wiss. Berlin (1959)

[Ne 49] von Neumann, J.: On rings of operators, Reduction theory, Ann. of Math. 50, 401 - 485 (1949)

[Ne 77] Neumann, M.: On the Strassen desintegration theorem, Archiv der Mathematik 29, 413 - 420 (1977)

[NMP 71] Nachbin, L., S.Machado and J.B.Prolla: Weighted approximation, vector fibrations and algebras of operators, J. Math. Pures Appl. 50, 299 - 323 (1971)

[Or 69] Orhon, M.: On the Hahn-Banach theorem for modules over C(S), J. London Math. Soc. (2) 1, 363 - 368 (1969)

[Pr 77] Prolla, J.B.: Approximation of vector valued functions, Mathematics Studies 25, North Holland (1977)

[Pr 79] Prolla, J.B.: The approximation property for Nachbin spaces, Approximation theory and functional analysis, Proc. Conference Approximation Theory, Campinas 1977, North-Holland Math. Studies 35, 371 - 382 (1979)

[Ri 74] Rieffel, M.A.: Induced representation of C^*-algebras, Advances in Mathmatics 13, No. 2, 176 - 257 (1974)

[Sc 72] Scott, D.: Continuous lattices, in: Toposes, algebraic geometry and logic, Lecture Notes in Mathematics 274, 97 - 136 (1972)

[Sch 71] Schaefer, H.H.: Topological vector spaces, Graduate Texts in Mathematics, Springer-Verlag (1971)

[Sch 77] Schaefer, H.H.: Banach lattices and positive operators, Grundlehren der math. Wissenschaften 215, Springer-Verlag (1977)

[Si 59] Singer, I.: Sur les applications linéaires intégrales des espaces des fonctions continues I, Rev. Math. Pures Appl. 4, 391 - 401 (1959)

[Sio 70] Sion, M.: A proof of the lifting theorem, University of British Columbia (1970)

[St 51] Steenrood, M.: The topology of fibre bundles, Princeton University Press, Princeton, New Jersey (1951)

[Su 69] Summers, W.H.: A representation theorem for the biequicontinuous completed tensor product of weighted spaces, Trans. Math. Soc. 146, 121 - 131 (1969)

[Va 75] Varela, J.: Sectional representation of Banach modules, Math. Z. 139, 55 - 61 (1975)

[We 65] Wells, J.: Bounded continuous vector valued functions on a locally compact space, Michigan Math. J. 12, 119 - 126 (1965)

[Wi 71] Wils, W.: The ideal center of partially ordered vector spaces, Acta Mathematica 127, 41 - 77 (1971)

INDEX

LIST OF SYMBOLS